HISTORY OF THE
ROYAL ASTRONOMIC.
VOLUME 2
1920 – 1980ˊ

CW00434926

The cover illustration is the emblem
of the Royal Astronomical Society
which contains its motto and the
40 ft telescope of Sir William Herschel,
the first President of the Society

HISTORY OF THE
ROYAL ASTRONOMICAL SOCIETY
VOLUME 2
1920 – 1980

EDITED BY

R.J. TAYLER
Professor of Astronomy
University of Sussex

WITH CHAPTERS BY
HIM AND BY

C.W. ALLEN
Emeritus Perren
Professor of Astronomy
University College London

Sir Wm McCREA FRS
President 1961–63
Emeritus Professor of Astronomy
University of Sussex

The Late H. DINGLE
President 1951–53
Emeritus Professor of
History and Philosophy of Science
Imperial College of Science and Technology

D.H. SADLER OBE
President 1967–69
Formerly Superintendent of
HM Nautical Almanac Office

G.J. WHITROW
Emeritus Professor
of Mathematics
Imperial College of Science
and Technology

FOREWORD BY

SIR HAROLD JEFFREYS FRS
President 1955–57
Emeritus Plumian Professor of
Astronomy and Experimental
Philosophy
University of Cambridge

PUBLISHED FOR THE
ROYAL ASTRONOMICAL SOCIETY BY
BLACKWELL SCIENTIFIC PUBLICATIONS
OXFORD LONDON EDINBURGH
BOSTON PALO ALTO MELBOURNE

© 1987 by The Royal Astronomical
Society and published for them by
Blackwell Scientific Publications
Editorial offices:
Osney Mead, Oxford OX2 0EL
  (*Orders*: Tel. 0865 240201)
8 John Street, London WC1N 2ES
23 Ainslie Place, Edinburgh EH3 6AJ
52 Beacon Street, Boston
  Massachusetts 02108, USA
667 Lytton Avenue, Palo Alto
  California 94301, USA
107 Barry Street, Carlton
  Victoria 3053, Australia

First published 1987

Set by Setrite Typesetters Ltd
Hong Kong
Printed and bound
in Great Britain
at the University Press,
Cambridge

DISTRIBUTORS

USA and Canada
  Blackwell Scientific Publications Inc
  P O Box 50009, Palo Alto
  California 94303
  (*Orders*: Tel. (415) 965-4081)

Australia
  Blackwell Scientific Publications
  (Australia) Pty Ltd
  107 Barry Street,
  Carlton, Victoria 3053
  (*Orders*: Tel. (03) 347 0300)

British Library
Cataloguing in Publication Data

Royal Astronomical Society
  History of the Royal Astronomical
  Society.
  Vol. 2: 1920–1980
  1. Royal Astronomical Society — History
  I. Title   II. Tayler, R.J.
  520′.6′041   QB1.R5
  ISBN 0-632-01791-0
  ISBN 0-632-01792-9 Pbk

Library of Congress
Cataloging-in-Publication Data

History of the Royal Astronomical Society.

  Vol. 2–   published : Oxford
[Oxfordshire]; Boston: Published for the
Society by Blackwell Scientific
Publications.
    Contents: [1] 1820–1920/edited by J.L.E.
Dreyer and H.H. Turner — v. 2. 1920–1980/
edited by R.J. Tayler.
    1. Royal Astronomical Society —
History.   2. Astronomy — England —
London — Societies, etc. — History.
I. Dreyer, J.L.E. (John Louis Emil),
1852–1926.   II. Turner, Herbert Hall,
1861–1930.   III. Tayler, R.J. (Roger John)
QB1.R75H57   1923   520.6041   24-13274

ISBN 0-632-01791-0
ISBN 0-632-01792-9 Pbk

CONTENTS

LIST OF PLATES

Plate 1. Sir Harold Jeffreys FRS, President 1955–57, Gold Medallist 1937. (Photograph taken in 1953 when Sir Harold was knighted, © A.C. Barrington-Brown.)

FOREWORD

I am glad to learn that the history of the Society for 1920 to 1980 is being written as I have been a Fellow since 1915, having been elected to a Fellowship at St John's College, Cambridge in November 1914; I compounded for subscriptions immediately and it was a good investment. I owe a great deal to the Society and in particular have welcomed the attention paid to geophysics. The history of this since the British Association Geophysical Committee first met in 1917, with H.H. Turner in the Chair and J.H. Jeans as Secretary, will be told by the authors. This became the RAS Geophysical Committee in 1919 and I succeeded Sydney Chapman as Secretary in 1920 until 1927.

Turner was active until his sudden death in 1930 while presiding at the meeting of the International Commission on Seismology of the IUGG at Stockholm. His discovery of deep-focus earthquakes was in a paper in volume 1 of the *Geophysical Supplement*. The Geophysical Discussions were usually reported in *The Observatory*.

I was at the meeting at which it was decided that women could be admitted as Fellows. Someone asked about the legal costs and he was told that it would be less than two composition fees. I did not remember that there were any votes against but apparently there were three.

My best wishes go to the Society for progress and success in the future.

Harold Jeffreys

INTRODUCTION

The first volume of the *History of the Royal Astronomical Society* was commissioned to celebrate the Centenary of the Society in 1920. The *History* was published in 1923 one year after the delayed centenary celebrations of 1922. In 1966 Council decided to publish a second volume of the *History* for the period 1920−70 in connection with the celebrations of the Sesquicentenary of the Society. Dr Alan Hunter was appointed as editor of the volume and it was decided that each of the five decades should be the responsibility of a single Fellow of the Society, which was also the original plan for the first volume. It was hoped that the second volume would be published in the early 1970s.

For various reasons it did not prove possible to keep to this timetable. In 1979 Council learnt that, although three of the chapters had been completed, the originally chosen author for the 1930s, Professor H.H. Plaskett, was unable to write his chapter. He was uniquely qualified to do so; from 1932 until the Second World War he probably did not miss a meeting of the Society, he was elected to the Council in 1933 and he was Secretary 1937−40 (and President 1945−47). Unfortunately failing health hindered his embarking on the task; about a year before his death in 1980 he had to inform the Society that it would be necessary to find someone else to undertake it. Council asked Professor W.H. McCrea to write the chapter and at the same time decided to extend the volume to include the years 1971−80. It is for this reason that the present volume covers the somewhat unusual period of 60 years. At a later date Dr Hunter asked to be relieved of the editorship. As a result his name does not appear on the title page of the present volume but I must acknowledge here the work which he put into the planning of the volume and to the commissioning of the original authors of the chapters.

There is one further modification to the original authorship. Professor H. Dingle was asked to cover the 1920s. He had served on Council from 1926 and he was Secretary 1929−33 so that he had an inside knowledge of the decade. He completed the draft of his chapter in 1970 but he died in 1978 before the whole of the manuscript of the volume had been assembled. When several of the present authors then read Dingle's draft, we found his coverage of the decade was very uneven and that one section was inappropriate for the *History*. As a result I

1

agreed to rewrite the chapter, incorporating much of what Dingle had written but adding supplementary material. The detective among our readers may be able to decide which sections of the chapter are in Dingle's original words.

Although each chapter in the volume covers approximately one decade there are detailed deviations from this and one important one. Chapter 1 covers the period from the Annual General Meeting in 1920, which marked the end of Volume 1, until the end of 1930. Chapter 2 takes the story up to the outbreak of World War II and Dr D.H. Sadler, who was invited to write about the 1940s, was asked to start his chapter by giving an account of the Society's actions taken in anticipation of the outbreak of war — which came in 1939 September — and to continue from there until the end of 1949. Chapter 4 runs from the beginning of 1950 until early 1960 and Chapter 5 finishes at the end of 1970. The final chapter covers the period 1971−80.

The authors were not originally given specific instructions regarding the length and contents of their chapters and as a result there is a significant variation in the size of the chapters. The chapters were written independently and no serious attempt has been made to provide cross-references between them. Where one happening in one decade had effects in subsequent decades, the event was to be described by the author for the first decade and the effects by the authors for the later decades. However, an author might say of something in his decade that nothing like it had happened before or since, or he might say of something else that the same has been happening ever since. Such statements could be true at the time of writing, but may not have remained true until the date of publication. An attempt has been made in final editing of the manuscript to see that there are no such incorrect statements in the published version.

The Index to this volume is constructed on the basis of that of the first volume. In particular, the name of an individual is indexed — in general with surname only — only in connection with some topic discussed in the text. It would be of little further value simply to index every occurrence of a name.

The period covered by this volume is one in which both astronomy and geophysics and the Society and its activities changed dramatically. Although the membership of the Society only increased by a factor of about 3, there was much larger proportionate increase in the number of professional astronomers and geophysicists, including a significant overseas membership. In 1920 a very large proportion of the Society's Fellowship consisted of amateurs and of scientists from other disciplines and they made important contributions to the Society's meetings and publications. It was common for such Fellows to be represented on Council and three amateur astronomers occupied the Presidential chair in the period covered by this volume, the latest being W.H. Steavenson in 1957–59. In contrast, by 1980 it had become very rare for someone other than a professional to speak at a meeting, publish a paper or be a member of Council.

The decline inside the Society of the amateur in the widest possible sense was no doubt related to the great increase in scientific knowledge. In 1920 astronomy was on the brink of absorbing ideas from the General Theory of Relativity and new discoveries from the 100-inch Mt Wilson telescope, but the generally accepted view remained that the Sun was near the centre of our stellar system, the Galaxy, which formed the whole Universe. The decades that followed saw the discovery of an expanding Universe of thousands of millions of galaxies containing many previously unknown types of object discovered by use of techniques across the entire electromagnetic spectrum from radio waves to gamma rays. In geophysics the most dramatic developments were in the acceptance of continental drift and plate tectonics and in spacecraft studies of the Earth's environment. The Society responded by greatly increasing the scope of its meetings and the size and contents of its publications. The meetings continued to be very well attended in spite of the large increase in the number of competing scientific meetings and the greater ease of foreign travel and they were frequently strengthened by the participation of overseas scientists. The dramatic discoveries in astronomy and geophysics and, in particular, those involving space research gripped the popular imagination and the present volume of the *History* ends with a largely professional Society, at least as far as its active membership is concerned, needing to make certain that it keeps in touch with this great popular interest in its subject.

This introduction is being written almost six years after the end of the period covered by this volume. Although it would not be appropriate to discuss in any detail what has happened in those six years, a few remarks are not perhaps out of place. The scope of the Society's activities has continued to grow but at the slower rate of the late 1970s rather than at the rapid rate of the previous one and a half decades. At the same time there has been serious concern, not only in the United Kingdom, about cutbacks in funding of universities and other institutes of higher education and about an increased emphasis on applied rather than pure science. Our science will need to exploit the great popular interest mentioned in the previous paragraph to help to press the case for further strong support for pure science. Sixty years ago the Society would have expected to have been consulted before any decisions were taken to build new telescopes or to change the organization of United Kingdom astronomy; indeed the Society initiated and pressed the case for the Isaac Newton telescope and was strongly represented on the Board of Visitors of the Royal Greenwich Observatory. Now the Society has to argue its right to express a view, as has been seen in recent discussions about the possible move of the Royal Greenwich Observatory from Herstmonceux.

All of the authors are grateful to successive Executive Secretaries in office at the times of writing their chapters, E.C. Rubidge, A.J. Hume and J.A. Steff-Langston, and Librarians, Mrs E. Lake and P.D. Hingley for their help in gathering together material for use in the *History*. In addition Sir William McCrea is grateful to the Royal Society for assistance from its grant-in-aid for the History of Science in carrying out some of the archival research necessitated by his enlistment to write about an early chapter. I am personally indebted to Professor McCrea for suggestions which have been incorporated in this Introduction and in Chapter 1. All of the authors are particularly pleased that Sir Harold Jeffreys agreed to write a foreword to the volume. He was a Fellow of the Society throughout the period and he was a member of Council for all but one year of the first decade and his subsequent Council service included the Presidency for 1955−57.

The plates in this volume have been chosen to illustrate events in the Society's history and the Society's possessions as well as major personalities during the period covered. The latter include two authors of chapters, Sir William McCrea and Dr D.H. Sadler. Professor Whitrow and I feel that their contributions to the well-being of the Society have been so substantial that the volume would not be complete without their photographs.

R.J. Tayler

CHAPTER 1

THE DECADE 1920–30

H. DINGLE & R.J. TAYLER

Introduction

The Royal Astronomical Society came into being in 1820 February in the aftermath of the Napoleonic wars at a time of social, industrial and economic transition. The Society began its second century of existence only a month after the coming into force of the Treaty of Versailles that brought to an end World War I, which had much greater consequences than the Napoleonic wars. It was a time of great international readjustment and again a time of great social and economic transition. The date of commencement of this chapter may be an accident of history but it was the start of a new epoch in world history. It happens also to have been a turning point in the science of astronomy and, indeed, of fundamental physics.

In 1919 November there was a joint meeting of the Society and the Royal Society, which is not described in the first volume of the *History*, but which will be discussed further below. The Astronomer Royal, Sir Frank Dyson, announced that the findings of the British eclipse expeditions of that year gave strong support to the predictions of Einstein's theory of general relativity concerning the 'bending' of light in the Sun's gravitational field. This theory, following upon the special theory of relativity, was producing a revolution in ideas about space and time, and it was yielding the beginnings of modern cosmology. Also in 1919 the 100-inch telescope at Mt Wilson was just becoming operational and by the end of the decade it was providing observations of distant galaxies which stimulated further developments in theoretical cosmology.

An even larger change in understanding of physical laws was brought about by the development of the quantum theory, with the introduction of the Heisenberg uncertainty principle. Particles not only existed in a curved spacetime but, at least at the microscopic level, their behaviour was no longer deterministic. Already in 1920 early developments in atomic physics were giving the first detailed understanding of physical conditions in the atmospheres and interiors of the Sun and the stars and were enabling the first accurate ideas to be obtained of the chemical compositions of objects, which could not be analysed directly by chemists. As the decade developed, the astrophysicist had to adapt to a succession of new quantum mechanical ideas, which were applied very rapidly to astronomical phenomena.

During the decade theoretical astrophysics became a well-established subject particularly with the work of Eddington, Jeans and Milne on stellar structure and of Eddington and Milne on stellar atmospheres. The meetings of the Society were enlivened by vigorous disputes about the structure of stars between Eddington and Jeans in the middle of the decade and between Eddington and Milne at the end. Some further comments will be made later but, from the accounts of the meetings of the Society in *The Observatory*, it is clear that the disputes were both hard hitting and good humoured. The publication of Eddington's book *The Internal Constitution of the Stars* in 1926 was a landmark in the subject, although not all would have agreed with this judgement. Looking ahead 20 years, Milne was still disagreeing with Eddington's approach in his 1945 Presidential Address entitled *The natural philosophy of stellar structure.*

As has already been mentioned, the main development in physics was that of the quantum theory in 1925-27. Many of the early disputes about the state of matter in stars arose because a proper quantum description was not available, but by the end of the decade the Society was given accounts by R.H. Fowler and Milne of the structure of white dwarfs based on the quantum statistics of a Fermi-Dirac gas. In addition quantum theory, in the form of Gamow's theory of nuclear barrier penetration, vindicated Eddington's belief that the centres of stars were hot enough for energy-releasing nuclear reactions to occur, although the elucidation of the detailed reaction chains had to wait another decade.

In observational astronomy there were equally important developments which were well covered in the Society's meetings and in the pages of *Monthly Notices*. The most dramatic change was from a Universe of one stellar system, with the Sun near its centre, to one of countless galaxies in an expanding Universe. In the early part of the period the position of the Sun in the Galaxy and the nature of the extragalactic nebulae were discussed frequently. For several years van Maanen's direct measurements of the rotation of the nebulae seemed to rule out an extragalactic position for them and many speakers at the Society's meetings gave strong support to van Maanen's views. However by the end of the decade the measurements had been discredited and the large distances to the nebulae and the expansion of the Universe were generally accepted. It was also clear that the Sun was far from the galactic centre. Another important development was the Oort-Lindblad account of the rotation of the Galaxy, which was supported by several papers presented to the Society.

The decade also marked a change in studies of stellar spectra and in the discussion of the chemical composition of stars and gaseous nebulae. Bohr's atomic theory had given a first insight into the existence of energy levels and of the transitions which produced spectral lines and this was reinforced by the more rigorous treatment of quantum theory. In addition Saha's discussion of ionization equilibrium had indicated that the temperature of stellar atmospheres and not the chemical composition of the star was the main factor influencing their spectra. As a result interpretation of stellar spectra was taking over from pure classification.

Of course these developments in astrophysics did not mean that everyone immediately stopped working in classical astronomy. The vital work on stellar motions and parallaxes continued and stimulated important theoretical work on stellar dynamics. Celestial mechanics was yielding further information about the motion of the Moon and planets and, at the end of the decade, the discovery of Pluto showed that the solar system still had secrets to disclose. The daily records of the Sun's activity taken at Greenwich would prove valuable to solar physicists many decades later. The amateur members of the Society continued to do extremely valuable work with their visual studies of the planets and their observations of comets, variable stars and novae.

The Society was very active in its support of geophysical meetings and it introduced a *Geophysical Supplement to Monthly Notices* so that geophysicists had a recognized outlet for their researches. Although the geophysical meetings brought some new people into the Society, there was, in fact, a very considerable overlap between the two subjects and those who worked in them. It is an interesting feature of the development of geophysics in Britain that it was closely linked with astronomy and that, as a result, astronomy and geophysics are still linked in one Society. Airy as Astronomer Royal started geomagnetic observations at Greenwich and this work continued to be a direct concern of the Royal Greenwich Observatory until it was transferred to the Natural Environment Research Council in 1965. In 1922 H.H. Turner, the Savilian Professor of Astronomy at Oxford, gave the first discussion of deep-focus earthquakes. Geophysics was developing rapidly and some highlights of the Society's activities during the decade were concerned with magnetic storms and the properties of the ionosphere, with seismic waves, with the measurement of gravity at sea, with ocean tides, with the first very detailed discussion of solar−terrestrial relations by Chapman and with the suggestion by Jeffreys that the Earth possessed a liquid iron core.

The Emergency Period

We must now pass to the second of the events that made the period around the year 1920 a juncture in the history of our Society as distinct from that of astronomy. Because of the greatly increased complexity of life in general, and of international relations in particular, the 1914–18 war produced a far more profound effect on matters non-political than did the Napoleonic wars, and accordingly our Society was presented with practical problems unknown to its founders and it is to these that we now turn.

Of these problems the most pressing was almost inevitably financial. All costs had risen, income which might have become available to the Society from increased membership and other sources had, of necessity, been diverted to the requirements of national survival, and the renewal of activities long suspended made financial demands which were not adequately met by the savings resulting from their suspension. In addition, the costs of the Centenary celebrations and the publication of the Society's *History*, which of course could not long be delayed presented imperative claims. Furthermore the Society's very efficient Assistant Secretary, Mr W.H. Wesley, had reached an age at which it had become necessary to appoint a permanent helper, and in 1920 March, Miss K. Williams, who had for some time given temporary service, was appointed Clerk at a salary of £130 per annum. On the death, late in 1922, of Mr Wesley, to whose 47 years of service deserved tribute was paid, Miss Williams was appointed Assistant Secretary. In spite of its financial problems, Council decided that it was necessary to continue to have a second member of office staff and Miss E. Wadsworth was appointed as Clerk.

These, and other, financial difficulties called for special measures and an 'Emergency period' was declared, during which these measures were to be decided upon and taken. One decision was that conditions were too abnormal for the Centenary to be celebrated in 1920 and it was ultimately celebrated in 1922 May as will be described later. It is perhaps an indication of how imperfectly the effects of the war were envisaged that, in response to an enquiry from the Treasurer concerning 'the exact meaning of the "Emergency period" and the date of its expiration', Council decided in 1920 January 'that the period be regarded as terminating on 1921 February 8, assuming peace with Germany to be concluded before 1920 February 8.'

Such unfounded optimism was, of course, not peculiar to our Society. On the contrary, we might even be considered rather far-sighted in establishing an 'Emergency period' at all. Towards the close of the first volume of the *History* (published in 1923), the writer notes the greatly increased cost of printing but adds (p.246), 'Still there is no need for immediate anxiety. In the past generous benefactors have occasionally by bequest increased the funds of the Society.' A footnote was inserted to the effect that 'Since this was written, the Society has, in 1922 April, received a liberal donation of £2500 from the Hon. Sir Charles Parsons, K.C.B., as a memorial to his father, William, third Earl of Rosse, the maker of the great telescope.' To this should be added the gift, by Mrs Harry Watson in 1920, of a sum of £300 in memory of her late husband, and a further bequest by her of £1000, which took effect in 1923, the total sum now being known as the Harry Watson Memorial Fund. The prospect of further benefactions was thus far from illusory.

At the same time, the full extent of the emergency became realized only gradually, and the steps needed to meet it still more gradually. The greatest embarrassment was the rapid increase in the cost of printing, and various measures were taken to deal with this problem. In 1920 April it was decided to ask the Conjoint Board of Scientific Societies (of which more presently) to 'approach the Government on the subject of some grant in aid to the various Scientific Societies, to meet the heavy expenses thrown upon them by the war.' The Board replied by submitting 'a series of questions' of which the only ones deemed by Council to be worthy of consideration were the first two, namely, 'Is your Society in favour of an application being made by the Conjoint Board to the Government for financial assistance to aid certain of the conjoint Societies?', (which was answered in the affirmative) and, 'If such an appeal is to be made, would your Society wish to enter a claim to be considered by Government?' (which inexplicably was answered in the negative).

A possible reduction in length of papers was next considered, and in 1920 November a small Publications Committee was appointed with 'plenary powers, without appeal to the Council' to curtail papers in the interests of economy. The raising of the Annual Contribution of Fellows was contemplated but apparently deemed impracticable even though it had not been changed since the foundation of the Society in 1820; this is in great contrast to reaction to the financial crisis of the 1970s, which is described in Chapter 6. To meet the cost of work connected with the Franklin-Adams charts, an overdraft of £560 was requested from the bank and granted, but, the general problem still remaining unsolved, a committee was appointed on 1921 May 13 'to consider and report upon the present state of the Society's finances.' Its practical proposals, however, amounted to little beyond stricter control of length of papers, dispensing with galley proofs, and discontinuance of the gift of 50 free reprints to authors (a privilege which was restored in 1930 February). A sidelight is thrown on the gulf between the nature of the problem and the Society's appreciation of it at that time by a Council minute of 1920 March 12 which, without further comment, states, 'A letter from the Chemical Society was read, stating that in the view of their Committee no useful purpose would be served at the present time by Scientific Societies attempting joint action to restrain printers from demanding further increases.' Comment is likewise needless now.

One man alone seems to have realized the drastic nature of the action needed — though, in acknowledging the uniqueness of his vision, it would be unjust to the Society as a whole not to acknowledge also its response to his initiative. Col. E.H. Grove-Hills, a contributor to the first volume of the *History* and past-President and Treasurer, whose bequest of an invaluable collection of early astronomical and mathematical books to our library is inadequately noted in the Preface to that work, and which will be discussed later in this chapter, saw clearly that the problem before the Society was far beyond solution by such trumpery measures as those so far adopted, and he took it upon himself to institute a special fund to meet the situation. A Council minute of 1922 February reads as follows:

> On the motion of the President, Col. Grove-Hills, who was nominated by the Council for election at the Annual General Meeting to the office of Treasurer, was invited to attend the meeting of Council. Col. Grove-Hills made a statement relating to the financial condition of the Society, and reported that, on his own responsibility, and without taking any action that committed the Council to accept his proposals, he was endeavouring to obtain a sum of money that would meet the special needs of the next four or five years. He estimated this sum at £1800. He had already collected, or received promises of support, as follows:-
>
> | | |
> |---|---:|
> | From certain Fellows | £500 |
> | From Fellows who had compounded before 1890 | £86 |
>
> The amount actually to be paid over that day was £535. 4s.0d.

Col. Grove-Hills had also written to Fellows, who had compounded their contributions under the new scale, to American Fellows, asking for their support and for their efforts in securing new Fellows, and to Associates. He had already received in reply many letters warmly approving his endeavour. The minutes of Council record:

> On the motion of Dr Crommelin, seconded by Professor Fowler, the Council approved the action of Col. Grove-Hills in initiating the collection of a special fund to tide the Society over the financial difficulties of the next four or five years.

and

> It was decided that Col. Grove-Hills should make a statement on the new financial situation at the Annual General Meeting. ·

He did so to good effect. It is sad to record that he died not many months later, on 1922 October 2, leaving to the Society the 'Grove-Hills Library' together with a sum of £250 to cover its removal to Burlington House, where it is now suitably placed in a special room, and having established the 'Grove-Hills Fund' on a sound basis. The total sum which it brought to the Society, including dividends received on investments during its course, was £1913.1s.11d. This was in remarkable agreement with his estimate of the amount needed, and it is unquestionable that it was his prompt and far-sighted action that enabled the Society to survive the post-war years without crippling, if not mortal, disaster. Having said that it is of some interest to compare the financial position of the Society in 1920 with that of 1945 after the Second World War. As will be seen from the figures in Appendix 5, the ratio of the financial reserves of the Society to its annual turnover was much the same in 1920 as in 1945 and the deficits on the Society's accounts immediately after the Second World War were relatively much worse than those after the First, but no similar emergency period was declared. This was perhaps related to different attitudes to science and scientists by the authorities in the two wars, which will be mentioned later, and to optimism after 1945 that support for astronomy and the number of astronomers would grow.

The membership of the Society was just under 800 in 1920 and it grew rapidly in the next three years to 900. This growth in membership at a time when many professional incomes were stationary or declining vindicated Council's decision not to increase the Annual Contribution. By the end of the decade the membership was 968. This proved to be a high point and there followed a decline, albeit a slight one, throughout the depression years of the 1930s and the figure of 1000 was not reached until 1949 . There were actually 1000 members at one stage in 1931 but this figure did not appear in the Table of Progress attached to the report of the Annual General Meeting because of losses of members before the end of the year.

The Centenary Celebrations

As has already been mentioned, because of the emergency Council decided that it would be inappropriate to celebrate the Centenary in 1920, and the celebrations were delayed until 1922, when they took place on several days in the week 1922 May 29 to June 3. Plate 2 is a photograph of those attending taken in front of the Society's premises. The first volume of the *History* was also produced to mark the Centenary and it appeared in 1923, but it contains no account of the activities of the Centenary because they fell outside its period. On the evening of May 29 a conversazione was held in the rooms of the Royal Society, kindly lent for the occasion, when astronomical exhibits were on display and lantern slides relating both to astronomy and to the history of the Society were shown. On the following day there was an all-day meeting of the Society. In opening the meeting the President (Professor Eddington) read the message which had been sent to His Majesty the King, Patron of the Society:

May it please Your Majesty!

We, the Fellows of the Royal Astronomical Society, assembled to celebrate the Centenary of our foundation, desire to send a message of homage and loyalty to Your Majesty, our Gracious Patron.

We look back on a wonderful century of astronomical discovery and we take pride in the contribution made by the Fellows of our Society, both to pure scientific knowledge, so essential to the progress of mankind, and to the valuable applications of astronomy in navigation and in the survey of the Earth.

From the far corners of the Empire we are banded together to keep a zealous watch on the skies, and to penetrate, as far as human limitations permit, the mysteries of the Universe.

We look forward to maintaining and extending these activities in the future under Your Majesty's continued encouragement and Patronage.

The President also also read the following reply:

Buckingham Palace

I am commanded by the King to thank you and the Fellows of the Royal Astronomical Society for your loyal message on the occasion of the Centenary of its foundation.

During the last hundred years the civilised world has witnessed a wonderful advancement in astronomical discovery, and His Majesty is proud to know that the Royal Astronomical Society both here and the far corners of the Empire has given its full measure of support towards the success of these valuable achievements.

You can rest assured that the King watches with interest and admiration the patient diligent and unobtrusive manner in which the Fellows of the Society conduct their unremitting research in the hope that they may, by piercing the hidden mysteries of the skies, add step by step to the store of scientific knowledge and thus contribute so much that is essential to the progress of mankind on land and sea.

STAMFORDHAM

Extracts of letters and telegrams from Associates and from astronomical societies and observatories were also read including one which epitomized the high regard felt for the Society:

American Astronomical Society sends greetings to her mother on hundredth anniversary.

In the morning session which followed, the President, Dr Dreyer and Professor Turner spoke about the history of the Society, about the personalities of some of its famous past members and about the developments of astronomy during the past century. In the afternoon six distinguished Associates, Professor F.H. Seares (Mt Wilson), Professor E. Strömgren (Copenhagen), Dr H. Shapley (Harvard), Professor E. Hertzsprung (Leiden), Dr R.G. Aitken (Lick) and Dr C.E. St John (Mt Wilson), spoke about some of the research being undertaken at their respective institutions. The day ended with a dinner at the Criterion

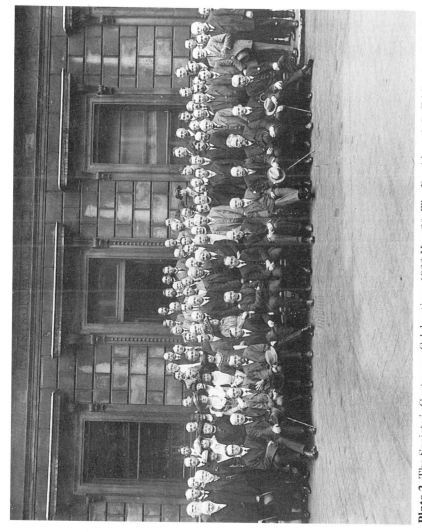

Plate 2. The Society's Centenary Celebrations 1922 May 30. The President A.S. Eddington is in the centre of the front row with the Secretaries T.E.R. Phillips and A.C.D. Crommelin on his right and left. The front row also includes foreign associates. (© Royal Astronomical Society.)

Restaurant attended by 120 Fellows and guests. The principal guests were Lord Balfour (former Prime Minister), Sir J.J. Thomson (President of the Royal Society) and the Presidents of several other scientific societies. Lord Balfour proposed the toast of the Society and replies were made by the President and by Dr St. John. Col. Grove-Hills proposed 'The Guests' and to this Sir J.J. Thomson and Professor Seward (President of the Geological Society) responded.

On the following day, May 31, members of the Society were invited to attend a meeting of the British Astronomical Association, which was designed to be part of the Centenary celebrations and where addresses were again given by five of the visiting Associates. Finally on Saturday June 3 Fellows and Associates were entertained at the Royal Observatory, Greenwich on the occasion of the annual visitation of the Observatory by the Board of Visitors.

The Centenary celebrations were a great success but earlier in the year one Fellow had suggested that it was not yet a suitable time for them. In the announcement of the arrangement for the meetings, which we have just described, there appeared the natural and apparently harmless statement, 'It is hoped that all Associates who find it possible will be present.' This prompted one senior Fellow to state his intention of moving, at the Annual General Meeting, 'That this Meeting repudiates the action of the Secretaries in expressing the hope that German Associates will be present at the celebration of the Centenary of the Society.' This was considered by the Council at some length on 1922 February 10. A suggestion that the Centenary celebrations should be postponed, presumably until anti-German feeling had somewhat abated, was rejected. Finally a minute was adopted recording that 'The Council, having considered the situation arising from the wording of the centenary circular, do not think that any formal action can be wisely taken.' Nevertheless, the proposed resolution was moved and seconded at the Annual General Meeting but it was not carried and no action was taken.

Officers of the Society

The Presidents

Six Presidents had at least part of their office in the decade, A. Fowler, A.S. Eddington, J.L.E. Dreyer, J.H. Jeans, T.E.R. Phillips and A.C.D. Crommelin. Two of the Presidents, Eddington and Jeans, were theoretical astrophysicists. Some of their contributions to astronomy and their disagreements are mentioned elsewhere in this chapter, but here it should be said that, in addition to their scientific distinction, they were both remarkable for their ability to convey their ideas both to their colleagues and to the general public. They both wrote several very well-received and influential textbooks which were published by the Cambridge University Press, with Jeans writing on electricity and magnetism and the dynamical theory of gases as well as on astronomy, while Eddington's *Mathematical Theory of Relativity* introduced many students to that subject. More than 60 years later the terms 'Jeans mass' and 'Eddington limit' are both on the lips of the avant-garde astronomer. Both Eddington and Jeans were knighted and were later appointed members of the Order of Merit. Many astronomers of future generations, including the junior author of this chapter, owe their first interest in astronomy to the popular writings of these two.

Fowler was a distinguished spectroscopist. He was the assistant and then successor of Sir Norman Lockyer at South Kensington what was one of the component parts of what ultimately became Imperial College. When Lockyer retired, the instruments were removed to Cambridge and Fowler became what would now be called a laboratory astrophysicist. The first accurate determination of the Rydberg constant and of the mass ratio of electron and proton were by-products of his work. Fowler was also the first General Secretary of the International Astronomical Union in 1920. He was first elected to Council in 1904 in unusual circumstances, which will be described later in this chapter. He was President in the Centenary year but, as has already been described, was unable to preside over the Centenary celebrations.

Dreyer was Danish by birth but he passed his astronomical career in Ireland, first as Astronomer to Lord Rosse at Birr Castle and afterwards as an assistant at Dunsink and as Director of the Armagh Observatory. His most substantial research was concerned with nebulae and he was the author of the *New General Catalogue* (NGC) which was published as a *Memoir* in 1888 and of two index catalogues published in *Memoirs* in 1895 and 1908. It is likely that most users of the NGC nomenclature today do not associate it with the name of Dreyer or realize that the catalogue was published as a *Memoir of the Royal Astronomical Society*. Dreyer was also well known as a historian of astronomy. He wrote *Life of Tycho Brahe* and *History of the planetary system from Thales to Kepler*. He edited the complete edition of Sir William Herschel's work on behalf of the Royal Society and the Royal Astronomical Society and he was joint editor of the first volume of the *History*. Whereas Fowler, Eddington and Jeans were in early middle age when they became President, Dreyer was at the end of his career. He was in retirement and he died in the year after the end of his Presidency.

The Reverend T.E.R. Phillips was a perfect example of the traditional member of the Society. In the early years there were few professional astronomers and many of the fellowship were amateurs, some of whom possessed telescopes and made observations and wrote papers and made a contribution to astronomy as great or greater than many professionals today. His chief subject of study was Jupiter. From 1900 to 1933 he was Director of the Jupiter section of the British Astronomical Association. The observations made by him and by younger colleagues such as F.J. Hargreaves and B.M. Peek provided the best available information concerning the surface features of Jupiter until the space missions of the 1970s. Phillips also made many observations and skilful drawings of Mars and he showed that long period variable stars could be divided into two groups by a harmonic analysis of their light curves.

Crommelin was a member of the staff of the Royal Observatory, Greenwich. He was Director of the comet section of the British Astronomical Association and he worked on the ephemerides of the Moon and the planets. He also wrote regular articles on discoveries of minor planets and comets. He was best known for his joint work with P.M. Cowell on the orbit of Halley's comet. In 1906 different predictions for the date of the next return varied by three years. Cowell and Crommelin produced an orbit which gave a date of perihelion which differed from the actual date by only three days and they also identified the past apparitions as far back as 240 BC.

The main duties of the President were to preside over the meetings of Council and of the Society. He was also expected to give a Presidential Address in each of the two years as President. Usually these addresses were associated with the award of the Gold Medal and were devoted to a discussion of the work of the medallist. Some of these addresses were remarkably detailed, even when their subject matter was remote from the personal research interests of the President, and they must have involved a considerable amount of work. Towards the end of the decade the institution of the George Darwin Lecture allowed overseas Gold Medallists an opportunity to describe their work in their own words, but it was only after the end of the decade that Presidents took the opportunity thereby presented to deliver addresses on subjects of their own choice. Only Dreyer delivered no full address on the award of a medal. His two medallists were both British, Eddington and Sir Frank Dyson, and it was probably thought that their work was already sufficiently familiar to members of the Society. Details of the Presidential Addresses can be found in Appendix 1.

Treasurers and Secretaries

At the start of the decade the Treasurer was E.B. Knobel. He had taken office in 1913 and had therefore seen the Society through the difficult years of the First World War. He was succeeded in 1922 by E.H. Grove-Hills, who died within the year having launched the special appeal for funds mentioned earlier. He was followed by F.J.M. Stratton (1923–27),

P.H. Hepburn (1927–28), Knobel again (1928–29) and then J.H. Reynolds started a term as Treasurer which would last, with just one intermission of two years while he was President, until 1946. This was an extremely large number of Treasurers in one decade; there were only nine Treasurers, one of whom, Knobel, served two terms, in the first century of the Society's existence. That the Society did not suffer from lack of continuity in this important office was largely due to the foundation laid by Grove-Hills.

In 1920 A.C.D. Crommelin and T.E.R. Philips were Secretaries and H.H. Turner was Foreign Secretary. Turner was Foreign Secretary until his death right at the end of the decade. He was well known for his 'From an Oxford notebook' in *The Observatory* and one only has to read his column to realize that he kept in very close touch with the world of astronomy and geophysics and that he must have been an ideal Foreign Secretary. He was also joint editor with Dreyer of the first volume of the *History*. H. Spencer Jones, J. Jackson and H. Knox Shaw also served as Secretaries and the decade ended with H. Dingle and W.M. Smart in post.

The Council

The composition of Council in the 1920s was very different from that of Council in the 1970s. Although there were requirements that the President and the Vice-Presidents could not serve for more than two years in the same capacity and that four of the ordinary members of Council must be new in that position each year, there was no reason why many of them could not remain on Council by changing their position. In fact each year there were typically two new members of Council who had not served on it before but otherwise there was great continuity. When J.W.L. Glaisher died in 1928, he was in the middle of his 55th consecutive year on Council. That was the record but Knobel also served 53 years with only one break of one year and Turner 43 successive years. Looking at the Council lists it is noticeable that they were largely composed of past, present or future Presidents of the Society, with the number in that category varying from 15 (out of 21) in 1924 to 19 in 1926. The number of different persons who were elected to Council in the 11 years 1920–30 was only 41 and of these only 15 never held the Presidency.

Of course the Council was not as homogeneous as the above description makes it sound. Some of the future Presidents such as Dingle, Jeffreys and Milne were in their late 20s or early 30s; Dingle and Jeffreys were not to become President until the 1950s and they were serving with other councillors 40 or 50 years older than themselves. Glaisher, who was President 1886–88, and Jeffreys, who was President

1955–57, were on Council together for the first eight years of the decade. What is clear is that within the total membership of the Society there was a very much smaller proportion of professionals than there is now and that election to Council carried with it a reasonable expectation of further advancement within the Society. Of the officers in the decade, only Hepburn, who was Treasurer for one year, did not achieve the Presidency and he died less than two years after his year as Treasurer. A reading of the pages of *Monthly Notices* reinforces the view of a rather small, professionally active element in the Society, as a very large fraction of the papers was authored by the same group of past and future Presidents.

Obviously there was a small group of Fellows who effectively controlled the Society and there were occasions when other Fellows were dissatisfied. It has already been mentioned that Fowler's first election to Council was unusual. The following extract from his Obituary Notice by H.C. Plummer published in *Monthly Notices* in 1941 tells the story:

> A curious interest attaches to Fowler's first election to Council of our Society, and there can be harm now in recalling the incident. At the end of the last century Council had tended to become something like a close corporation and had lost contact with a considerable body of Fellows. Members of Council were commonly re-elected year after year, and when a new choice became necessary (as a rule by death) the additional candidate was usually found in a narrow clique. It is true that a remedy was provided then, as it still is, by the clause which allows any Fellow to substitute another name for any one printed in the Council's list. But in practice the clause is apt to be ineffective, because it needs a sufficient number of Fellows to write not merely in support of one candidate but against one particular official nominee. At last, in 1904, the necessary combination was successfully organised, and an unofficial candidate was duly elected.

In that manner Council acquired a future Secretary, President and Foreign Secretary. At the time that he was President, the clique was not so tight but at the end of the decade, as will be mentioned later, Council took action to ensure a more regular turnover of Council membership.

Clearly the members of Council knew one another very well. Most of them were also members of the Royal Astronomical Society Club, so that they will also frequently have dined together on the Ordinary Meeting days. Clearly this close-knit nature of Council made for an efficient handling of business but it also led to some inertia in decisions as will be mentioned subsequently.

Obituaries

Nine past Presidents of the Society died during the decade. They represented a very wide spectrum of experience, which was common in

Fellows of the Society in the early years of this century but which has now largely been lost because of the very great increase in the number of professional astronomers and in the breadth and depth of scientific knowledge.

Sir W. de W. Abney, who died in 1920, was President in 1893–95. His early career was spent in the Royal Engineers and he later served in the Science Department of South Kensington, being successively Inspector of Local Schools and Principal Assistant Secretary, Science Branch of the Board of Education. He was particularly distinguished for his researches in colour and colour photography and much of the success of the three-colour photographic process resulted from his work.

Sir William H.M. Christie, who died in 1922, became Chief Assistant at the Royal Observatory, Greenwich in 1870. He joined the Society in 1871 and served on Council from 1872 until 1912, apart from 1879–80. He was Secretary in 1880–82 and President in 1888–90. He succeeded Sir George Airy as Astronomer Royal in 1881. Airy had been very economical in his running of Greenwich and Christie had a considerable task to modernize the buildings and other facilities. Once he became Astronomer Royal he took little practical part in observing except in some eclipse expeditions but concentrated on being an efficient organizer of the work of others. He must be particularly remembered as the founder of *The Observatory* in 1877.

Much has already been said about E.H. Grove-Hills, who also died in 1922. He was Treasurer in 1905–13 and President in 1913–15, both under the name of E.H. Hills, before he added his mother's maiden name. He was a professional soldier who was director of the geophysics section of the General Staff at the War Office during the Boer War. He was very much concerned with both theoretical and practical geodesy. He was also secretary of the committee which determined the boundary between Chile and Argentina when their governments appealed to King Edward VII as arbitrator. His main astronomical interest was in solar physics and for many years he was secretary of the Joint Permanent Eclipse Committee.

W.H. Maw, who died in 1924, was an engineer by profession. He was originally involved with railways and he designed locomotives for the East Indian Railway. He was one of the editors of the journal *Engineering* from its foundation in 1865 until his death in 1924. He was a very active consulting engineer and at different times was President of both the Institution of Mechanical Engineers and the Institution of Civil Engineers. He devoted his leisure to astronomy, building his own observatory and making micrometer measures of variable stars. He was a Council member of the Society from 1892 to 1919 and was Treasurer in 1900–05 and President in 1905–07.

The career of J.L.E. Dreyer has already been described, as has the very long Council service of J.W.L. Glaisher, who died in 1928. He was a very distinguished pure mathematician at Trinity College, Cambridge and an inspiring teacher. He was for many years an editor of the *Quarterly Journal of Mathematics* and of the *Messenger of Mathematics*. He was Secretary in 1877–84 and President in 1886–88 and in 1901–03. He was also President of the Royal Astronomical Society Club for 33 years.

P.A. MacMahon, who died in 1929, was another President who was a distinguished pure mathematician, whose major publications included work in number theory and combinatorial analysis. He was not, however, a mathematician by profession. He was a member of the Royal Artillery, who became Professor of Physics at the Ordnance College. After retirement he became Deputy Warden of Standards at the Board of Trade and General Secretary of the British Association in 1903–13. He was President in 1917–19.

E.B. Knobel's long Council service has also been mentioned. He was an analytical chemist who was successively Manager and Head Brewer at Messrs Bass and Co., Burton on Trent, Manager of Courtauld's silk factory and Managing Director of Ilford Photographic Laboratories. He was a very keen amateur astronomer from childhood and he was particularly interested in oriental astronomy. He translated the Ulugh Beg catalogue of stars and he edited the star catalogue part of the translation of Ptolomy's *Almagest*. He made some particularly fine drawings of Mars at the 1884 opposition. He was Secretary in 1882–92, Treasurer in 1895–1900, 1913–22 and 1928–29 and President in 1892–93 and 1900–01.

Both Knobel and H.H. Turner died in 1930. Turner was Chief Assistant at Greenwich in 1884–93, when he became Savilian Professor of Astronomy at Oxford, a post which he held until he died. At Oxford he directed very important and laborious work on the astrographic catalogue. He had a personal research interest in variable stars and he discovered Nova Gem (1903). His geophysical researches, including his work on deep-focus earthquake waves, have already been mentioned. It was he who was the originator of a meeting held in 1917 to consider means of promoting discussion and publication of geophysical investigations, which led, as will be explained below, to the Society's taking some responsibility for geophysics. He was Secretary in 1892–99, President in 1903–05 and Foreign Secretary from 1919 until his death.

Although other distinguished Fellows and also Associates died during the decade, the above brief outlines of the careers of the deceased Presidents give an excellent idea of the range of interests and expertise represented in the Society, which neither then nor now was a purely professional institution. It is also clear that even in the 1920s a substantial move had already been made towards the time when most

Council members would be professional astronomers and geophysicists; this is a reflection of the new era in astronomy brought in by general relativity and the 100 inch telescope.

The Gold Medal

A major routine item of Council business was the award of the Gold Medal of the Society. In 1920, for the first time since 1891, no Gold Medal was awarded. In normal circumstances this would not have been the case and it would have been awarded to A. Einstein. As a result of the historic meeting in 1919 November, which has already been mentioned, the meeting of the Society in the following month had been wholly devoted to relativity and it had been attended by leading workers in astronomy and theoretical physics in this country. The report of their addresses shows an overwhelming but not unanimous acceptance of the theory as a permanent contribution of outstanding importance to astronomical knowledge. 'It was a fitting tribute to the great theory', says Volume 1 of the *History* (p.238), 'which had been raised from the rank of an interesting hypothesis to that of an epoch-marking theory by the confirmation afforded it by experiment and observation.'

But in the early 1920s circumstances other than astronomical were far from normal. Einstein was a German and technically England was still at war with Germany. The award of the Society's highest honour to a citizen of an enemy country was therefore not a matter that could be decided on astronomical grounds alone, and it is not surprising that the relative weights of the various considerations were not equally estimated by all members of Council. Although Einstein was selected for the award of the Gold Medal in 1919 December, when only a majority vote was required, the award was not confirmed in January when a vote of three-quarters in favour was needed. In fact neither the original proposer of Einstein's nomination, Turner, nor the seconder, Jeans, was present in January and there might have been a different result if they had been there; Council voting figures are not recorded. Eddington committed the grave error of telling a Quaker friend who was visiting Germany after the December meeting that he could tell Einstein that he would get the medal and he had subsequently to write a very apologetic letter to Einstein. In 1920 November when, according to routine, the award for 1921 came up for discussion, Einstein, among others, was proposed, but opinion was too much divided to make his selection possible. The controversy was repeated in the following years, and it was not until 1926 that astronomical claims were felt to be sufficiently free from political interference to allow the award to be made to Einstein unanimously and without undesirable consequences.

Two things must be borne in mind by those venturing to pass judgement on the Society's treatment of this matter. In the first place, the award of the medal rests wholly with the Council, whose decision cannot be questioned by the Fellows in general. Second, in those days, as has been mentioned earlier, the composition of the Council changed much more slowly than the present rules allow. A number of persons of special standing in astronomy were, in effect, permanent members, for when their permitted term of office in one capacity (e.g. Vice-President) expired, they could be re-elected in another role, as ordinary members, say. The influx of new members was much smaller and opposition to the strong nucleus of VIPs (to use a modern colloquialism which would then have been regarded with horror) would have required great strength of character and in all probability would have been ineffective. The consequence was that the balance of judgement on a point tended to remain constant, at least until there was sufficient change in public feeling to reverse it. Such a long delay in awarding the Gold Medal to Einstein therefore cannot be considered a failure of the Society to acknowledge to the full his unique contribution to astronomy. It must also be recognized that it was perfectly possible for a member of Council to approve of Einstein and his work but to feel that the time was not yet right for the award.

The Gold Medal was awarded in all of the other years of the decade, to three Americans (H.N. Russell, A.A. Michelson and F. Schlesinger), to four Britons (J.H. Jeans, A.S. Eddington, Sir Frank Dyson and R.A. Sampson), to one Canadian (J.S. Plaskett) and to one Dane, resident in the Netherlands (E. Hertzsprung), in addition to Einstein. The work of most of these Gold Medallists needs no introduction even in the 1980s as all astronomers have heard of the Hertzsprung–Russell diagram, the Michelson–Morley experiment and the Michelson interferometer and of the many contributions to theoretical astrophysics by Jeans and Eddington.

The citation for the award to Eddington mentioned his researches in star streaming, the internal constitution of the stars and generalized relativity, while the award to Jeans was primarily for his work in cosmogony. Russell was honoured for his work on stellar evolution, while, in the case of Hertzsprung, his determination of the distance to the Lesser Magellanic Cloud was particularly mentioned in addition to other pioneering researches in stellar astronomy. The remaining ci-tations were: Michelson, 'Application of interferometry to astronomical measurement'; Dyson, 'Astronomy in general and in particular for his work on the proper motions of the stars'; Sampson, 'Theory of the four great satellites of Jupiter'; Schlesinger, 'Stellar parallax and astronomical photography'; Plaskett, 'Valuable observations of stellar radial velocities and the important conclusions derived from them'. Overall, the awards represented a very good mix of traditional and avant-garde astronomy.

It is of interest to look back through the minute books of Council to see who was not awarded the Gold Medal despite being proposed regularly. The most notable case in this decade was Miss A.J. Cannon, who was proposed no fewer than nine times. Another distinguished candidate who was proposed several times but never received the medal was Dr M.N. Saha.

Returning to Einstein, it is a pleasure to add that, during his visit to England to lecture on his theory, Einstein, accompanied by Freundlich (then a supporter of the theory of which he later became strongly critical), attended the meeting of the Society on 1921 June 10. This was, appropriately, during the Presidency of Eddington who, using the new language appositely on this occasion, remarked on the special interest in the fact that the world-lines of Einstein and the Royal Astronomical Society should cross. Einstein was given a warm welcome and he thanked and complimented the Society on its work. He spoke in German, which was translated by Lord Haldane.

Two other items relating to Gold Medallists are worthy of mention. In 1925 Council discovered, to its embarrassment, that it had for some time been failing to observe one of the provisions of the will of Sir John Herschel which said that a copy of his *Cape Observations* should be presented to each Gold Medallist. Council ascertained that there were 31 copies remaining in stock and agreed that a copy should be offered to each future medallist until all copies were exhausted. A further embarrassment occurred in 1930. A report was received that Michelson had died and the Ordinary Meeting stood in his memory. Fortunately the information that he was, in fact, alive and well reached the Society before the Foreign Secretary had sent the Society's condolences!

The George Darwin Lecture

A now firmly established feature of the Society's activities, the George Darwin Lectureship, originated during the decade. At that time the Society had no special lectureships other than the Presidential Address, which was in fact usually an address on the award of the Gold Medal. When he was President, Dr (later Sir) J.H. Jeans decided that it would be to the Society's advantage to possess a named lectureship. At the Council meeting of 1926 June 11 the following letter from him was read.

June 6, 1926

My dear Treasurer,

The Royal Astronomical Society has no lectureship or other endowment such as makes it possible, in many Societies, for an eminent person from abroad to be invited to give a lecture on his own work. It would give me great pleasure if the Council should see fit to accept the following offer:

I offer the sum of £1000 (a thousand pounds) for the endowment of a 'Foreign Lectureship', the capital to be kept intact, and the income to be used for an annual lecture on some subject of interest to astronomers, preference being given to a lecturer normally resident outside the British Isles. I had thought that in particular the medallist of the year, when resident abroad, might frequently be invited to be Foreign Lecturer also:- it would give him an extra inducement to come in person to receive his medal. When no suitable Foreign Lecturer was available, Council could of course invite one of their own Fellows or someone in some cognate science, Geophysics, or even Physics of the right kind.

If Council accepts this offer and if, after fair trial of the Foreign Lectureship, they consider the money could be used better in other ways, I should of course wish this to be done. I should prefer (but without creating any legal obligation) that the income should be used for some specific purpose and not merged in the general funds of the Society, but quite realize this might not be possible.

Will you be so good as to bring this matter before Council?

Yours sincerely
J.H. JEANS.

This offer was accepted 'with gratitude for its munificence, and admiration for its thoughtfulness', and a small committee was appointed to arrange details with the President. As a result a form of Trust Deed was drawn up and sealed, and the following recommendations were accepted:

(1) That the Treasurer's action in investing the gift in 5% War Stock be confirmed.

(2) That the Annual Lecturer for any calendar year shall ordinarily be nominated at the preceding December Council after the selection of the nominee for the Gold Medal, and be elected at the January meeting in each year.

(3) That in the event of his declining, the Council shall at the same later meeting select another person to lecture in the same calendar year.

(4) That the lecture be given at an Ordinary Meeting or at a Special General Meeting called for the purpose, but preferably not at the Annual General Meeting.

It was further decided:

That the lectureship be called the George Darwin Lectureship, in honour of a great astronomer, but that this name be not taken to imply restriction of subject to a special department of astronomy.

Thus came into being a feature of the Society's work, like a source of progress and adornment of its activities, which continues amply to fulfil the wishes and expectations of its creator. Eventually the existence of the lectureship led to the main Presidential Address frequently being a subject of the President's own choice unrelated to the award of a medal but this did not occur in this decade. The first four lectures in 1927−30 were delivered by F. Schlesinger, W.H. Wright, E. Hertzsprung and J.S. Plaskett. The full list of George Darwin Lecturers is given in Appendix 4.

Geophysics in the Society

It is now appropriate to record the development of the Society's concern with geophysics as a legitimate branch of astronomy. From an extraterrestrial viewpoint, the Earth is merely one of a number of satellites of the Sun whose cosmic status is in no way influenced by the circumstances that, since it is our dwelling-place, we can study it by means not immediately applicable to its sister satellites. The Society had always regarded the Earth as a proper subject of study and its earlier members included the ill-fated arctic explorer Sir John Franklin and the surveyor Sir George Everest. There were, however, very important developments just before and during the decade. As Sir Harold Jeffreys relates in his Foreword to this volume, the British Association for the Advancement of Science set up a Geophysical Committee in 1917 and this became a Royal Astronomical Society Geophysical Committee in 1919, so that the Society was subsequently regarded as the learned society with prime concern for geophysics.

It is not out of place here to record a suggestion by one Fellow, S. Chapman, which did not receive the support which perhaps it deserved. He proposed that the various Earth Sciences, geography, geology, geodesy, geophysics, etc., should receive a single name, for which he suggested geonomy. It is not out of place because of a current regrettable tendency to speak of the sudden development of selenology as the beginning of 'lunar geology'; this is as felicitous a designation as 'solar apogee' would have been for aphelion when heliocentric displaced geocentric astronomy. If this tendency is not checked, it will lead in due time to what might well at once be called 'planetology' or 'planetonomy' as including 'neptunian geology' and 'plutonic geology', terms which already exist with quite different meanings.

However, in the period with which this chapter is concerned the subject of interest was the proper designation of the Earth as a planet and not the improper designation of the other planets as the Earth. The Society's earlier concern with the subject is recorded in the first volume of the *History*, and the main advances in the decade as far as the Society specifically was concerned were the establishment of the

Geophysical Supplement to Monthly Notices, which was started by a resolution of the Council in 1922 April, and the holding of meetings specially arranged for the discussion of geophysical subjects. Before the publication of the *Geophysical Supplement*, papers on geophysical subjects had for some time appeared in the *Monthly Notices*, undistinguished from those on other branches of astronomy, but the new *Geophysical Supplement* gave geophysicists a focus for the publication of their researches. The special meetings made an unfortunate start, the first being called for 1926 May 7, which turned out to be in the midst of the General Strike, with the result that only one of the four papers announced could be presented. Nevertheless, despite slow beginnings, the attention paid to geophysics within the Society has grown continuously. A summary of the whole movement from its origin up to 1936 is given in *Monthly Notices*, volume **96**, p.384: the transformation of the *Geophysical Supplement* into the *Geophysical Journal* is recorded on p. 149 of this volume.

Having suggested that geophysics had a harmonious growth in the Society, it is also clear that there were at times some tensions between astronomers and geophysicists within the Society, even though some of the most prominent geophysicists, such as Turner, were also astronomers. One obvious problem was that the great growth of scientific knowledge which has continued to the present day was preventing individuals from taking an intelligent interest in the whole of both sciences. In 1928 March Council received and considered a memorandum on geophysics in the Society from a small committee headed by Jeffreys. It made four recommendations:

(a)   the *Geophysical Supplement* should be continued;
(b)   a Geophysical Section of the Society should be set up;
(c)   geophysicists should pay a reduced Annual Contribution;
(d)   geophysicists should be eligible to become Associates.

Of these four points Council accepted only (a). It is easy to see why it did not accept (b) and (c) as this would have caused divisions in the Society which might be repeated as other specializations emerged. In contrast the refusal to accept (d) looks grudging as the denial of (b) and (c) implies that geophysicists should be regarded as full members of the Society and particularly as Jeans had explicitly suggested that the George Darwin Lecture might from time to time be on a geophysical topic.

The Society's Meetings

We have already mentioned briefly the joint meeting of the Society and the Royal Society held in 1919 November to discuss the results of the eclipse expeditions of that year. These had been organized, as such expeditions continued to be organized for many years to come, by the Joint Permanent Eclipse Committee of the two societies. This meeting is, strictly, outside our period but it was not discussed in the previous volume of the *History*. The President of the Royal Society, Sir J.J. Thomson, was in the chair and the main part of the meeting was devoted to accounts by the Astronomer Royal (Sir Frank Dyson) and by A.S. Eddington of the eclipse expeditions to Sobral in Brazil and to Principe off the West coast of Africa, and of the results obtained which, within their estimated errors, were in full agreement with the prediction for the bending of light in Einstein's generalized relativity and which disagreed completely with calculations based on Newtonian gravitation and special relativity. There was general acceptance of this interpretation which supported the earlier success of generalized relativity in explaining the residuals in the motion of the perihelion of Mercury, although one speaker, Dr Silberstein dissented. Professor F.A. Lindemann (later Lord Cherwell) reported a pilot experiment which aimed at detecting the deflection of infrared radiation by the Sun outside eclipses.

Other meetings in the decade that followed were to receive a variety of contradictory results by observers such as J. Evershed concerning the attempt to measure the so-called third crucial test of general relativity, the gravitational red shift of spectral lines.

The general atmosphere of the time is represented by a quotation from a book by A.N. Whitehead (*Science and the Modern World*, Cambridge University Press, 1926, p.15):

> It was my good fortune to be present at the meeting of the Royal Society in London when the Astronomer Royal for England announced that the photographic plates of the famous eclipse, as measured by his colleagues in Greenwich Observatory, had verified the prediction of Einstein that rays of light are bent as they pass in the neighbourhood of the sun. The whole atmosphere of tense interest was exactly that of the Greek drama: we were the chorus commenting on the decree of destiny as disclosed in the development of a supreme incident. There was dramatic quality in the very staging: the traditional ceremonial, in the background the picture of Newton to remind us that the greatest of scientific generalisations was now, after more than two centuries, to receive its first modification. Nor was the personal interest wanting: a great adventure in thought had at length come safe to shore.

In an earlier publication (*The Concept of Nature*, Cambridge University Press, 1920, p.182) Whitehead commented on the problems which would be faced by physicists in studying the new subject:

> It is not going too far to say that the announcement that physicists would have in future to study the theory of tensors created a veritable panic among them when the verification of Einstein's predictions was first announced.

Although there were soon some good accounts of the theory, including Eddington's *Mathematical Theory of Relativity*, there continued for a long time to be the myth that general relativity was far beyond most scientists. The view was summed up in the presumably apocryphal exchange, 'Professor Eddington, is it true that only three people understand general relativity?' 'Oh, who is the third?' The typical astronomer of the 1920s would be surprised to find the subject commonly taught in undergraduate courses of the 1980s.

Council of the Society decided that its own meeting on 1919 December 12 should also be devoted to a discussion of general relativity and this meeting is described briefly in the first volume of the *History*. There were introductory talks by Eddington and Jeans and a communication from Sir Joseph Larmor was read in his absence. Dissent was again expressed by Silberstein and by Sir Oliver Lodge, who was unwilling to give up the aether. Lindemann said that he foresaw some problems with boundary conditions when an attempt was made to apply general relativity to the whole Universe.

An account has already been given of the special meetings associated with the Centenary. Another special meeting of the Society was held on 1925 July 24. The second General Assembly of the International Astronomical Union was held at Cambridge on July 14–22. This meant that a large number of foreign astronomers, including many Associates were in England and Council decided to hold a Special General Meeting which they could attend. The President, J.H. Jeans, presided and the speakers were Dr W.W. Campbell, Dr B. Lindblad, Professor H. Shapley, Dr H. Spencer Jones, Dr A. van Maanen, Professor J. Stebbins, Dr J.S. Plaskett and Professor W. de Sitter.

Campbell spoke about the work done at the Lick Observatory and at its southern station in Chile on the radial velocities of bright stars and of the implications of the results for ideas about stellar evolution. Lindblad discussed the determination of spectrographic absolute magnitudes and the distances of faint stars arguing that useful work could be done with objective prisms of low dispersion. Shapley spoke about some recent work which suggested that obscuring nebulosity, so evident in low galactic latitudes, was also present in high galactic latitudes. This was very important as it affected all work on the distribution of faint stars, clusters and nebulae. Spencer Jones discussed the discovery and subsequent changes in brightness and spectrum of 1925 Nova Pictoris. Van Maanen spoke about possible errors in his measurements of internal motions and rotation in spiral nebulae and argued that the results were sound. Stebbins described the photoelectric photometer which had been constructed at Madison observatory and he showed some light curves of variable stars obtained with it. Plaskett spoke about the spectroscopic work of the Victoria Observatory and in particular of the studies of stellar radial velocities, while de Sitter described the current work of the Leiden Observatory including proper motions, double stars, variable stars and work on the stellar reference frame.

Another notable meeting of the Society was the Ordinary Meeting of 1930 March 14, when news had just been received of the discovery of another planet in the solar system, later of course to be named Pluto. The President, A.C.D. Crommelin announced:

> Yesterday was the 149th anniversary of the discovery of Uranus by Sir William Herschel and the anniversary was marked in a very appropriate way by our receiving news of the discovery of another primary planet. I will read the cablegram which Professor Shapley with great courtesy has sent to the Society.
>
> 'Lowell Observatory Flagstaff search for Trans-Neptunian revealed object of the 15th magnitude. It has been observed for seven weeks. Both in the rate of motion and in its path it conforms to the approximate distance assigned by the late Professor Lowell. On the 12th of March at 3 hours its position was 7 arc seconds west of Delta Geminorum agreeing with Lowell's longitude. Shapley.'

The President announced that the Foreign Secretary had sent the following telegram:

> President, Council and Fellows of the Royal Astronomical Society in meeting assembled send to Lowell Observatory their hearty congratulations on the great discovery of Trans-Neptunian planet. Turner Foreign Secretary.

In the following year, at the beginning of the next decade, C. Tombaugh, who actually discovered Pluto, was awarded the Jackson Gwilt Medal and Gift of the Society.

The Ordinary Meetings of the Society throughout the decade were held in the early evening from 5 pm to 7 pm. Apart from the Presidential Address, which was in fact given at the Annual General Meeting which at that time coincided with the Anniversary Meeting in February, and in the latter part of the decade the George Darwin Lecture, they were almost entirely devoted to the reading of papers presented to the Society. These papers were subsequently printed very rapidly in *Monthly Notices*, which continued to be largely in effect a record, *in extenso*, of the monthly meetings. The discussion which followed the papers was reported, as it still is, in *The Observatory*.

It was in this decade that the celebrated debates between Jeans and Eddington regarding all the basic issues of stellar constitution made the Society's meetings so very famous. Other eminent astronomers were to a greater or less extent professionally involved and were often present; these included, for instance, S. Chapman, F.W. Dyson, R.H. Fowler, H.

Jeffreys, E.A. Milne and R.A. Sampson. Then there were other well-known scientists who came along in order simply to enjoy the spectacle, in much the same spirit as that in which some of them might watch a Test Match. It is often said that the great pure mathematician G.H. Hardy became a Fellow just so that he could share in the enjoyment. No doubt this is partly true; but it is also true that R.H. Fowler, who was a distinguished pure mathematician before he turned to mathematical physics, had made Hardy genuinely interested in the mathematical aspects of the debates. Although Eddington and Jeans disagreed vigorously at the Society's meetings, they were on good personal terms and this is illustrated by Plate 3 which shows them at dinner at the Royal Astronomical Society Club talking to a club guest, Mr S.C. Roberts of the Cambridge University Press, which published many of their books.

Towards the end of the decade these debates were still further invigorated by the intervention of E.A. Milne. Until then, his notable astrophysical contributions, nearly all of which had been published by the Society, had concerned stellar atmospheres and stellar spectra. Without much warning but with some flourish of excitement in 1929 he broke into the subject of stellar interiors as well. A main topic of disagreement between Eddington and Jeans related to the former's mass-luminosity relation. Eddington claimed to be able to obtain such a relation, which was satisfied by many of the stars whose masses were known, without having any knowledge of the mechanism of energy release in stellar interiors but Jeans disagreed strongly with this viewpoint. In 1929 November, Milne accepted that there was an observed correlation between luminosity and mass but denied that any theoretical argument could give this result because atomic physics was not sufficiently advanced to permit a fundamental calculation of the luminosity of stars. In particular he argued that an addition of mass at the surface of a star would not increase the star's luminosity.

In effect Milne was proposing the rejection of all the main results of Eddington's work, which were enshrined in his *Internal Constitution of the Stars*, which was already a much respected classic. Immediately after Milne's talk, Jeans said that he was glad that Milne had come round to his view regarding the independence of mass and luminosity. Eddington, in a lengthy reply, said that Milne had not given enough details of his work for it to be clear why his results disagreed with Eddington's, and 'my interest in the rest of the paper is dimmed because it would be absurd to pretend that I think that there is the remotest chance of its being right.' Eddington followed this up with a paper of his own at the 1930 January meeting in which he argued that Milne's view that the surface boundary condition had an important effect on the gross properties of a star was completely incorrect.

The intervention by Milne ensured that those meetings at which stellar constitution was a main topic reached a properly dramatic climax at the end of the decade. Thereafter no other single topic played such a dominant role. Inevitably the 1920s will go down in history as the decade of stellar constitution with Jeans and Eddington in star parts, and Milne providing a nova outburst at the end. Eddington's extraordinary physical intuition regarding the mass-luminosity relation was largely vindicated when nuclear energy sources, involving fusion reactions which were highly temperature sensitive, were discovered. In contrast Jeans had earlier supported radioactivity which would be essentially independent of the physical state of the material. Milne's own most important lasting work on stellar structure was that relating to degenerate stars which he first described in a paper to the Society on 1930 November 14.

National Affairs

The First World War and its aftermath led to some interaction between scientific societies and government. The authorities realized, if only to a very limited extent, that scientific research had some relevance to military operations, so that, for example, a man like the distinguished physicist H.G.J. Moseley might have been more effectively employed than in being killed in the Dardanelles. It is mentioned in the first volume of the *History* (p.227) that in 1916 the Society had agreed to join the newly formed Conjoint Board of Scientific Societies. This was established originally 'for the purpose of promoting cooperation in appealing to the Government on matters relating to science, industry and education', but inevitably the experience gained in the war pointed to the need for an extension of this function. Accordingly, in 1920 April the Conjoint Board approached the Society, as well, it is to be presumed, as all the other societies belonging to it, 'urging the advisability of approaching the Admiralty and the War Office on the subject of Science in relation to warfare, and asking for the comment or criticism of the Council.' A committee was appointed to draft a reply. In view of the now unquestioned role of science in warfare, it is of interest to record the report of the committee at that early date; it was transmitted to the Board without modification in the following form:

> The Council of the Royal Astronomical Society have given careful consideration to your letter of April 6, and venture to make the following suggestions as suitable for embodiment in any letter that the Conjoint Board may forward to the Admiralty and the War Office.
>
> It is admitted generally that in any future war the closest possible cooperation between scientists and the heads of the fighting forces is absolutely essential. This can only be obtained efficiently if Senior Staff

Plate 3. Royal Astronomical Society Club Dinner 1935 May 10: from left S.C. Roberts, Sir James Jeans and Sir Arthur Eddington. (Photograph by G. Merton. © Royal Astronomical Society Club.)

[*facing page* 30]

Officers and leading scientists are brought into fairly close contact in peace time, so that each side may learn to understand the other's point of view. For example — it might be possible to arrange for lectures to be given to each Staff course of the Army, Navy, and Air Force by scientists of established reputation. Again it might be possible to arrange some scheme by which problems confronting the fighting force could be put before scientists. In any case personal knowledge and acquaintance in advance would be a great help in preparing the close liaison which must be established, in the event of any future war, immediately after the outbreak of hostilities.

It would seem to the Council that the Army Education Scheme should allow for the refresher courses for regimental officers of the Army, which would deal generally with some special subject, such as chemical warfare, Physics in War or Psychology of War. Further, selected officers from the Staff and technical services might with advantage be sent to work in University laboratories or other scientific institutions. It might also be possible to arrange for the almost equally important visits of scientists to the Naval and military establishments with a view to their learning something of the actual war conditions and problems.

1. For two years after the outbreak of war the scientists employed at the Royal Aircraft Establishment at Farnborough made repeated applications to be allowed to learn to fly, but it was only in the autumn of 1916 that permits were granted.

2. As instancing the difficulty of persuading the authorities to take up new methods we may mention that the method of sound ranging afterwards adopted was put before the War Office in January 1915 by a Fellow of this Society, who received the reply that the method was quite impossible, and would not be proceeded with.

Again it was suggested early in 1916, and repeatedly urged that submerged objects could be detected more easily if the reflected light from the surface of the water were removed by means of some polarising device. It was not until 1918 that some hundreds of these were finally ordered.

3. On the other hand, there are many cases where a fuller acquaintance with the practical conditions of warfare would save scientists from making suggestions which could not possibly be adopted. To name two very flagrant examples: — it was suggested that the height of an aeroplane above the surface of the earth could be measured by the variation of the gravitational acceleration. If the originator of this scheme had ever been in an aeroplane he would have realised how ridiculous it was. Again, inventions were submitted and actually tried which involved using selenium cells on the wing-tips of miniature aeroplanes. These were to act on the controls when only one wing-tip was illuminated, and by this means it was hoped to keep the aeroplane in the beam of a searchlight. It is difficult enough to use selenium cells in a laboratory, and to suggest using them on an aeroplane shows a lack of knowledge of flying conditions only equalled by the ignorance of the controlability [sic] of an aeroplane evinced by the idea that a machine could be kept in the beam even if the cells worked.'

It would appear that the Conjoint Board failed to fulfil whatever expectations had led to its formation, for there is no evidence that it achieved anything of value, and much that it did not. Dissatisfaction grew, and at the meeting of the Board in 1921 December the question of the utility of its continued existence was raised and deferred until the following March. In the meantime, E.B. Knobel, one of the Society's two representatives on the Board, reported to the Council of the Society that he and his colleague agreed 'that the activities of the Conjoint Board had been productive of very limited results. Several schemes on which Committees had been appointed had proved abortive, and it was difficult to see what useful purpose remained for continuing the organisation.' The Council agreed that the continuance of the Board was unnecessary. Notwithstanding that this view was taken by several other societies, the Board decided that it should continue to exist, though meeting only once a year (except in emergency) on grounds of expense; whereupon the Institutions of Mechanical Engineers and of Mining and Metallurgy and the Royal Society of Medicine withdrew from it. In view of these facts the Council decided that the Society should withdraw also. It does not appear that this decision gave any cause for later regret. Although the Conjoint Board did not achieve any great success, it must be remarked that, as far as the use of scientists in wartime was concerned, the lesson of the First World War was learnt. Much more effective use was made of scientists in the Second World War and our own science and the Society benefited greatly, particularly from the development of radio astronomy out of wartime radar and from generally increased government support for science.

This decade was favoured by nature with a rare event, a total eclipse of the Sun visible in northern England and Wales on 1927 June 29. Realizing its responsibility to the public, the Society approached one of the main railway companies as to the possibility of running special trains to places within the zone of totality. The reply was at first negative and somewhat satirical but it changed later, as a consequence of press publicity, to a request for information concerning the degree of support which such an undertaking might receive. A serious proposal was made that a special LNER train should be run from London to Yorkshire but this was eventually dropped. At this stage W.J.S. Lockyer suggested to Council, of which he was a member, that the Air Ministry might be willing to make the airship R33 available to the Society. Although Lockyer believed the airship to be free, this proved not to be true. There was great public interest in the eclipse but adverse weather conditions limited the value of the observations from the point of view of astronomical progress and deprived many spectators of a unique sight.

A more favourable initial response came from the railway companies to an approach from the Society on the advisability of adopting the 24-hour day system in railway timetables. This is a subject on which the Society appears to have been more than a third of a century ahead of its time. In 1928 April the Council appointed a subcommittee consisting of the President (T.E.R. Phillips), the Astronomer Royal (Sir Frank Dyson) and the Foreign Secretary (H.H. Turner) to approach railway companies 'in the name of the Council with a view to promoting the adoption of the twenty four hour day in railway timetables.' The approach was made, and the companies expressed their willingness to introduce the system if there was a popular demand for it. A letter, signed by members of the subcommittee, was thereupon sent to the press. It appeared in *The Times* of 1928 December 8 and was accompanied by a leading article.

The letter requested the opinions of readers on the proposal. This stimulated some correspondence, the tenor of which was summarized in a further letter from the subcommittee which appeared in *The Times* on 1929 February 5. It pointed out that 20 correspondents favoured the introduction of the 24-hour system in railway timetables and 6 opposed it, and added, 'we suggest that a substantial case for change has been made out.' At the February meeting of the Council, shortly after this, it was decided to approach the railway companies again in the light of this information, and on March 20 Lord Lamington unsuccessfully moved in the House of Lords that the Government should commit the railway companies to this system. The companies were again approached by the subcommittee but this time they replied that they did not see their way to such a change until it was adopted by the country as a whole. The Council thereupon decided, at the 1929 June meeting, that the matter could not usefully be pursued further at that time. In fact, the 24-hour system was adopted in railway timetables in 1965 June.

Man, according to Shakespeare's Hamlet, is a being of 'large discourse, looking before and after'. Not only in the magnitude of its discourse, but also in the matter of time measurement, the Society was conspicuously human, for, having looked forward to the divisions of the day, it turned its gaze backwards to the beginnings of accurate chronometry. As is well known, a pioneer in the construction of artificial clocks was John Harrison, who in the eighteenth century was awarded a large monetary prize for devising and constructing such an instrument for the accurate determination of longitude. The Society possessed a clock made by Harrison, apparently for his own use, which, notwithstanding its age, was still not beyond repair. It had been examined and was described in the *Monthly Notices* for 1909 November by Mr E.T. Cottingham and had been partly adjusted by him. Early in 1927, Commander R.T. Gould, a man of wide interests, knowledge and ability, though not a Fellow of the Society, offered to make such further improvements as to restore it as nearly as possible to its original state and to enable it, if kept in an airtight case, to function for many years to come. The offer was gratefully accepted, and the work was successfully completed. The clock, in the proposed case, was installed early in 1929 in the Society's rooms at Burlington House, after a suggestion that it should be lent to the Science Museum for more general exhibition had been rejected. Later it was transferred to the Royal Greenwich Observatory and now it is in the National Maritime Museum, where it remains one of the Society's most valuable possessions (see Plate 14).

International Affairs

A proposal was made in 1921 that responsibility for the Cape Observatory in South Africa should be transferred from the Admiralty to the Union Government. The opinion of Council was sought and in its representation it said that, if such a transfer did take place, there were three essential conditions which must be met. The first was that the Observatory should be assured of adequate funds, the second that the publication of results should be guaranteed and the third that there should be a first class director. It has to be remembered that at that time there were very few observatories in the Southern Hemisphere so that observations taken at the Cape were of particular importance to the world of astronomy. In fact the proposal was dropped and the Cape Observatory remained the responsibility of the Admiralty until 1965 when it was placed under the newly formed Science Research Council. On 1972 January 1 it was incorporated in the new South African Astronomical Observatory. In fact until that date the history of the Observatory had paralleled that of the Society as the ordinance inaugurating the Royal Observatory of the Cape of Good Hope was signed on 1820 October 20 not long after the foundation of the Society and they both celebrated their Sesquicentenary in 1970.

Comment has already been made in several places about problems relating to the restoration of normal relations with German astronomers. A cognate, though less emotionally charged, problem to that presented by Germany arose in connection with intercourse with the Soviet Union. In 1924 April a proposal was received from the promoters of a scheme 'for renewing and creating intercourse between British and Russian scientists,' asking if the Society would co-operate formally with a society for this purpose if it were formed. After some consideration it was decided that personal co-operation with Russian scientists should be favoured whenever occasion might arise, but formal association of the Society with the proposed new body was not approved, though some exchanges of publications were arranged.

On several occasions Council was asked to aid astronomers overseas, who for one reason or another lacked adequate access to astronomical literature. In 1921 April, following an appeal from Serbia, Council agreed to offer to donate some duplicate library books. Knobel objected in a letter to the President that the Council had no power to give away the Society's property without permission of a General Meeting but the President ruled otherwise. Council agreed in 1923 November to donate journals to the Tokyo Observatory Library, which had been destroyed in the Tokyo earthquake.

The Grove-Hills Library

The Grove-Hills Library was bequeathed to the Society by its late Treasurer and ex-President and the bequest was mentioned briefly in the preface to the first volume of the *History*, although it fell outside the period covered by that volume. This very generous bequest of more than 500 books together with a sum of £250 to enable them to be housed properly provided the Society with what must be counted as its most valuable individual possessions. All but three of the volumes in the library were printed before the year 1700 and 40 had been printed earlier than 1501. A few of the more important and particularly interesting books in the library are:

Hyginus. *Poeticon Astronomicon*. Venice, E. Radolt, 1482.
An early printed version of this popular astronomy book written two millenia ago, with some of the first illustrations of the constellation figures.
Copernicus. *De Revolutionibus*. Nuremberg, 1543.
The foundation of modern astronomy: the first edition.
Galileo. *Istoria...Macchie Solari*. Rome, 1613.
Galileo declares the spots to be part of the Sun itself, and attacks the entire doctrine of Aristotle.
Hevelius. *Selenographia*. Danzig, 1647.
This great amateur astronomer prepared the first accurate atlas of the Moon from observations with his telescopes.
Tycho Brahe. *Historia Coelestis*. Augsburg, 1666.
Flamsteed. *Historia Coelestis Libri Duo*. London, 1712.
The 'unauthorised' edition published against Flamsteed's wishes: he burned as many copies as he could lay his hands on.

Grove-Hills collected these books at a time when interest in the history of science was very much less than it is now and the Society is fortunate to have such a collection which is accessible to the historian of astronomy.

Popularization of Astronomy

The very important developments in astronomy during the decade awakened a public interest in the subject which was wider than that of the traditional amateur astronomer. It is interesting to note the reactions of Council, which were very cautious. Although in 1923 May Council had discussions about inadequacies in the Science Museum's astronomical collection, which showed that it was concerned about the public face of astronomy, at the same time it declined to collaborate with the broadcasting authorities in arranging astronomical addresses by recognized experts. Again in 1924 June Council declined to help the British Association and British Science Guild to find a suitable lecturer on popular astronomy. In 1926 December a small committee was set up to consider a BBC letter about a course of astronomy lectures but again there is no record in Council minutes of any subsequent action. In 1929 no member of Council was willing to talk to Richmond Rotary Club. However, also in 1929/30 Dingle (Secretary) represented the Society at a conference called to appoint a committee of enquiry into questions arising from the rapidly increasing use of films, records and other visual and auxiliary devices in educational work, and in the same year Council agreed to assist the Royal Photographic Society in an exhibition of astronomical photographs.

What seems surprising in retrospect is that prominent in the Council during the decade were Eddington and Jeans, who played such a major role in the subsequent popularization of astronomy. Most British astronomers who started their careers in the following quarter century must have been strongly stimulated by their writings. Indeed *The Nature of the Physical World* by Eddington was published in 1928 and *The Universe Around Us* and *The Mysterious Universe* by Jeans in 1929 and 1930 respectively, so that they were both active in the propagation of astronomy to a wider public. *The Universe Around Us* was based on some wireless talks by Jeans so possibly the earlier approach to the Council by the BBC did have some ultimate effect.

Somewhat related to the question of popularization of astronomy was the problem which Council faced in both 1923 and 1928 when people who were not Fellows in good standing were practising as astrologers and describing themselves as FRAS. On the latter occasion Council seriously considered taking legal action but ultimately decided that this would probably do more harm than good. This view was almost certainly correct.

The Society's Staff

Mention has already been made of the death of W.H. Wesley, the Assistant Secretary, in 1922. He had taken up his post in 1875 February a bare three months after the Society's move from Somerset House to Burlington House and he died in service on 1922 October 22 at the age of 81. He was originally a wood engraver by profession and after joining the Society he made considerable use of his artistic talents in the service of astronomy. He engraved Dr Boedicker's charts of the Milky Way, which were published in 1892, and another major work was his map of the inner portion of the Moon's surface, which occupied all of his spare time, including holidays, from 1911 to 1915. He originally believed that photography could not compete with drawings of celestial objects but in 1900 he admitted that good photographs could be capable of showing all the details which can be distinguished by eye at the telescope.

After Wesley's death Council set up a committee to consider applications for the post of Assistant Secretary and it instructed the committee to select one name for Council to consider along with that of Miss Williams the Clerk. Council's decision was to appoint Miss Williams. This was not to the satisfaction of one of the other candidates, who was not the one who was selected for the final stage. He was a Fellow who objected strongly to the appointment of a female Assistant Secretary and he resigned his fellowship in protest. He was one of a very small minority. Miss Williams, together with Miss E.M. Wadsworth, who was shortly afterwards appointed Clerk, served the Society very well for the remainder of the decade and for a considerable time afterwards. In fact Miss Wadsworth was still alive as a pensioner of the Society when the Society entered the 1980s at the end of this volume of its history.

The Bye-Laws of the Society

The record of an institution such as the Society must present features of two different types. In the first place there are changes arising from unforeseeable events occurring both in the development of astronomy and geophysics and in the systematic and accidental changes in the general conditions of the world in which it operates. Second the Society's internal mechanism must be able to react to these changes and to perceptions, gained from experience, of possible improvements. The account so far is mainly concerned with the first of these features: it remains to depict the second.

The Society must work within the framework of its two Charters, whose requirements are absolute. It has considerable freedom within that framework to formulate bye-laws for the continuous performance of its functions, and it is obviously desirable, in order to maintain the stability of the Society, that these shall be expressed so as to permit the maximum possible freedom of action to meet changes, not of a fundamental character, without the formality of frequent amendment. The early history of the Society is a tribute to the wisdom of its founders and their successors in this respect. The general framework of the bye-laws remained largely identical with those originally drawn up until a major revision in 1977 which is described in Chapter 6. Evidence of both the scope of the first bye-laws and reluctance to change them without necessity is afforded by an application from the Geographical Section of the War Office early in 1924 for 'Institutional Membership' of the Society. The desire of the Council to accede to this request was insufficient to overcome its hesitation to make the necessary changes in the bye-laws, and a compromise was reached in the following resolution:

> That an Institution, on the payment of the ordinary annual subscription, be eligible on the nomination of the Council to receive the publications of the Society, and to send a representative to the ordinary meetings, but that such a representative shall not possess the rights of a Fellow.

This regulation continued to operate until 1978 when Council decided that Institutional Membership simply gave bodies the right to acquire the Society's journals at well below the usual library price and that no corresponding advantage was gained by the Society from the existence of the Institutional Members.

Notwithstanding all this, it is inevitable that changes both within and outside the Society will necessitate changes in the Society's bye-laws and the first major response to this necessity for many years occurred during the decade under discussion. It has already been mentioned that the semi-permanent constitution of the Council was one factor in delaying the award of the Gold Medal to Einstein, and this was probably one of a number of incidents that revealed the desirability of ensuring a freer introduction of new blood into the Council in order to maintain adaptation to the changing environment. Even as late as the beginning of our period, as we may see from the first volume of the *History* (p.245), the balance between 'experience' and 'stagnation' was regarded as satisfactorily preserved by a system which had allowed certain individuals to serve continuously for half a century or more. But the demands of post-war conditions could no longer be resisted and accordingly at the Annual General Meeting of 1928 February the Council put forward a number of changes of bye-laws, arising from certain 'principles' which were stated thus:

> (a) The List of Fellows nominated to serve on the Council to contain the names of at least four Fellows not serving on the existing Council;
> (b) Election of New Fellows to be made by the Council and confirmed by the Fellows at the Ordinary Meeting;
> (c) Election of Associates to be similar to that of Fellows;
> (d) Alteration of beginning of financial year from February 8 to January 1;
> (e) Restriction of amount of arrears allowed;
> (f) Minor changes.

It is evident that, while the composition of the Council was the main problem that led to these proposals, the undesirability of making frequent changes of bye-laws led to a more general survey of the subject, and the opportunity was taken to make all essential amendments at once. A revised method of presenting the annual statement of accounts had already been adopted in 1923 February to meet the increasing complexity of the Society's financial arrangements, but at that time it was felt necessary to work within the existing bye-laws, despite some inconvenience. The change of financial year would appear to have been proposed to remove that inconvenience.

CHAPTER 2

THE DECADE 1931–39

SIR WILLIAM H. McCREA

The Decade of the Thirties

The Society prizes both its Royal Charter and the fact that every Sovereign since King William IV has been pleased to become its Patron. The years concerned here saw an unusual number of royal occasions, some joyous, some sad and at least one of high drama. They included in 1935 the Jubilee of King George V, in 1936 the death of King George V, the Accession of King Edward VIII, his Abdication followed by the Accession of King George VI. As appropriate, the Society presented Loyal Addresses, and these are all recorded in its *Monthly Notices*.

They were unsettled times. At home there was the economic crisis of 1931 resulting in the National Government that continued in office until 1935. In Germany the National Socialists came to power in 1933 with consequences that are only too well known. In Italy the Facists were in power; they concluded a military alliance with Germany in 1939. In Spain the Civil War of 1936–37 resulted in a dictatorship. Then in 1939 came the apparently inevitable outbreak of World War II.

Until early 1939 when the Council began to consider 'the protection of the Society's property in the event of War', the Society's official records contain no single explicit mention that any one had noticed any of these momentous happenings, save for some reference to the fact that the war in Spain made it difficult for the Director of the Madrid Observatory to transmit his annual contribution as a Fellow. This is less surprising than might appear at first sight, because any economic troubles were of such a sort as to have little direct effect upon the Society, although sadly they made it impossible for a number of individuals to afford to remain Fellows. A world war is manifestly an altogether different matter. However, it is patently evident to those organizing this volume that the history of the Society's war years should be dealt with as a whole. It is fortunate that that history has been written by Dr Sadler who, as Secretary from 1939 to 1947, along with the late J.H. Reynolds, as Treasurer from 1937 to 1946 (and previously Treasurer 1929–35, President 1935–37), played an immeasurably greater part than any other individual in leading the Society through the whole of that unbelievably exacting time. Consequently he has been able to recount the story from unique knowledge.

Plate 4. Sir William H. McCrea FRS, President 1961–63, Treasurer 1976–79, Secretary 1946–49, Foreign Correspondent 1968–71, Gold Medallist 1976. (Presidential portrait, © Royal Astronomical Society.)

[*facing page* 40]

There is, however, one obvious drawback in having the story told by Dr Sadler: the chapter will not pay tribute to his own unequalled service to the Society. The only remedy appears to be for the author of the chapter leading up to Dr Sadler's, who happens to have first-hand knowledge of at any rate part of that service, to write these few words about it. Throughout the years concerned Dr Sadler was carrying major national professional responsibilities. Nevertheless he maintained intimate involvement in the Society's affairs; above all he ensured that its programme of scientific meetings and publications proceeded to the utmost extent that all the exigencies of the times could be made to admit. Thanks to his inspired leadership in these regards, when peace returned, the Society's activities recovered full vigour with astonishing smoothness and rapidity.

In consequence of all this, the present chapter covers nothing after the outbreak of war in 1939 September, and nothing of the action taken earlier in anticipation of this, except where these are inseparable from other affairs. With this qualification, it is still convenient to call the interval actually dealt with 'the decade'. Also when one writes in the sense that something used to be the case, this is intended necessarily to apply only during that decade — unless it is obvious that anything else is meant.

The war undoubtedly had a retrospective effect upon the record of the decade or thereabouts leading up to it. If an individual suffers some traumatic experience, he may lose the memory of what has happened to him not only for some time since the shock but also for a comparable time before it. Something similar seems to apply in the collective experience of a community. In particular, in consequence of the shock of World War II the memory of the scientific community for what happened during several years previous to it seems to be impaired in comparison with its memory for still earlier years. Everybody seems to recognize the greatness of the 1920s when quantum mechanics was discovered, when the internal constitution of the stars began to be understood, when the expansion of the universe was revealed, and so on, but surely the 1930s were no less remarkable than any decade before or since. It was the true beginning of the nuclear age, when new elementary particles were discovered (positron, neutron) others hypothesized (neutrino, meson/muon), when the immemorial dream of the transmutation of chemical elements was first realized, and the first artificial radioactive nuclei produced, when neutron stars were first discussed and when the source of stellar energy was revealed, showing indeed how the whole physical universe works. The first new window was opened upon the heavens by the discovery (in the USA) of radio emission from the Galaxy. Furthermore much of optical astronomy was revolutionized by the completion of the first Schmidt telescope. This is a book about a scientific society, not about a science, but the scientific background of the time is vitally relevant.

The Society was recognized as having strong international interests and relations. But naturally its roots were in British astronomy, and so something must be said about that as it was in the 1930s.

On the observational side, the Yapp 36-inch reflector came into use at Greenwich, and did good service in work on stellar spectra. The Radcliffe 74-inch reflector was being planned for erection at Pretoria; unfortunately much time was lost as a result of an unhappy controversy in which the Society never became in any way involved. The British amateur astronomer J.P.M. Prentice discovered Nova Herculis in 1934 and this became (chiefly at Cambridge and Greenwich) the most observed star ever, apart from the Sun. It was a time that saw the peak of solar astronomy so far this century; solar terrestrial relations were studied with much vigour at Greenwich and Cambridge, and the astronomy of the Sun itself with the new purpose-built tower telescope at Oxford. The solar total eclipse of 1936 became the most observed eclipse ever, and British expeditions had a large share in making it so.

On the theoretical side, work on stellar structure continued from the previous decade with undiminished intensity; again a major share was done in this country. The Milne–Eddington controversy about the appropriate solutions to the basic differential equations was in spate at the start of the decade. Interest then shifted to the structure of white dwarf stars and the Chandrasekhar–Eddington controversy about relativistic degeneracy, which was destined to affect the history of the subject for three or four more decades. Then towards the end of the 1930s interest focused upon stellar energy generation.

The end of the preceding decade had seen the discovery, by Hubble in California, of the expanding universe. Much of the resulting theoretical work came to be done in Britain, where several offshoots attracted considerable attention, e.g. 'Newtonian cosmology' and Milne's kinematic relativity.

It was almost certainly in Britain that *geophysics* first came to be recognized as a coherent science in its own right. The Society had much to do with this, particularly in the previous decade. However, the 1929 edition of the *Concise Oxford Dictionary* still did not mention the word! The decade under review was a time of expansion and consolidation in basic geophysics. It saw the publication of D. Brunt, *Physical and dynamical meteorology* (Cambridge, 1934) and S. Chapman and J. Bartels, *Geomagnetism* (2 volumes, Oxford, 1940), which joined H. Jeffreys *The Earth* (Cambridge, first edition, 1924) as probably the first great classics in the field. All were by prominent Fellows of the Society.

The holders of the main official astronomical offices in Britain were: Astronomer Royal, F.W. Dyson 1910–33, H. Spencer Jones

1933–55; Superintendent of HM Nautical Almanac Office L.J. Comrie 1931–36, D.H. Sadler 1936–71; Astronomer Royal for Scotland, R.A. Sampson 1910–37; W.M.H. Greaves 1938–55; HM Astronomer at the Cape of Good Hope H. Spencer Jones 1924–33, J. Jackson 1933–50; Savilian Professor of Astronomy and Director of the Oxford University Observatory H.H. Plaskett 1932–60; Plumian Professor of Astronomy and Director of the Cambridge Observatory A.S. Eddington 1913/14–44; Professor of Astrophysics and Director of the Solar Physics Observatory, Cambridge F.J.M. Stratton 1928–47.

In reading the formal records of the Society one might be persuaded into believing that the individuals whose names appeared so persistently had nothing more to do than to discuss the wording of bye-laws, the expenditure of bequests, the holding of soirées, and so forth. But it has always to be remembered that these same persons were the leaders in much of the massive scientific activity just mentioned and they included all these holders of greatly demanding scientific offices.

The Society in the Decade

The decade was a time of robust maturity in the existence of the Society. It is surely no mere coincidence that this also describes all those who filled the office of President in those years. All the Society's affairs were conducted with vigour; but thanks in no small measure to the personalities of these Presidents, it was a time too of remarkable harmony amongst all who worked with them in any capacity. It was indeed a time of economic stringency calling for continual prudence in material doings. Nevertheless the Society comported itself with proper style, vitality and enterprise — and unselfconsciousness — befitting the world's premier astronomical body.

It seems appropriate to begin by listing those who held office in the Society in any part of the decade.

Presidents

1929–31 A.C.D.Crommelin (1865–1939) Assistant at the Royal Observatory Greenwich 1891–1927; led the Greenwich 1919 eclipse expedition to Brazil to test the Einstein deflection of light past the Sun (Secretary RAS 1917–23).

1931–33 H. Knox-Shaw (1885–1970) Radcliffe Observer, Oxford 1924–39, Director Radcliffe Observatory, Pretoria 1939–50 (Secretary RAS 1926–30). He conducted the meetings with good humour and frequent flashes of wit.

1933–35 F.J.M. Stratton DSO, OBE, TD (FRS) (1881–1960) Professor of Astrophysics and Director Solar Physics Observatory, Cambridge

1928−1947; General Secretary IAU 1925−1938 (Treasurer RAS 1923 −27; Foreign Secretary RAS 1945−55). In the chair he showed a special talent for putting an inexperienced speaker at ease.

1935−37 J.H. Reynolds (1874−1949) Birmingham industrialist and distinguished amateur astronomer, since whom only one other amateur has held the office. (Treasurer RAS 1929−35, 1937−46). His great services as Treasurer throughout World War II are recorded in the chapter following this.

1937−39 (Sir) Harold Spencer Jones (KBE) FRS (1890−1960) Astronomer Royal 1933−55, President IAU 1945−48 (Secretary RAS 1923 −24; Treasurer RAS 1946−52; Foreign Secretary RAS 1955−60; Gold Medallist 1943). He was of impressive presence and presided over the Society's proceedings with notable dignity. Plate 5 is his Presidential portrait.

1939−41 H.C. Plummer FRS (1875−1946) Royal Astronomer for Ireland 1912−21, Professor of Mathematics, Military College of Science, Woolwich 1921−40. Without holding any other main office, he served the Society in many ways, not least in agreeing to write numerous obituary notices for *Monthly Notices*, in which he manifested kindly insight.

Between them, in addition to their personal qualifications for the office, these Presidents covered a wide range of the British astronomy of the time. Lest it be remarked that the range could have been even wider, it should be recalled that both Eddington and Jeans had been President in the 1920s and that the President in 1941−43 was Chapman, the first geophysicist to be elected.

Treasurers

1929−35, 1937−46 J.H. Reynolds (President RAS 1935−37).

1935−37 Sir Frank W. Dyson KBE FRS (1868−1939) Astronomer Royal 1910−23 (Secretary RAS 1899−1905; President RAS 1911−13). He took office for this interval because the Council wished Reynolds to serve a term as President.

Secretaries

The appointment noted is that held while serving as Secretary.

1929−33 H. Dingle (1890−1978) Assistant Professor of Astrophysics, Imperial College (President RAS 1951−53).

1930−37 W.M. Smart (1889−1975) John Couch Adams Astronomer, Cambridge (President RAS 1949−51).

1933−39 W.M.H. Greaves (1897−1955) Chief Assistant, Royal Observatory (President RAS 1947−49).

Plate 5. Sir Harold Spencer Jones FRS, President 1937–39, Treasurer 1946–52, Secretary 1923–24, Foreign Secretary 1955–60, Gold Medallist 1943. (Presidential portrait, © Royal Astronomical Society.)

1937–40 H.H. Plaskett (1893–1980) Savilian Professor of Astronomy, Oxford (President RAS 1945–47).

1939–47 (Acting Secretary 1938) D.H. Sadler (1908–   ) Superintendent HM Nautical Almanac Office (President RAS 1967–69).

1940–41 R.d'E. Atkinson (1898–1982) Chief Assistant, Royal Observatory.

On average Secretaries have served over four years each, while a President's term has always been limited to no more than two years. Consequently, even though at any time there have been two Secretaries in this category, and even though a few Presidents have served more than one term, somewhat more individuals have served as President than as Secretary. Up to the time of writing, about half the Secretaries have later become President, so in this respect the record for the decade is above average.

Foreign Secretaries

1931–36 A. Fowler CBE, FRS (1868–1940) (President RAS 1919–21).

1936–44 Sir A.S. Eddington OM FRS (1882–1944) (President RAS 1921–23).

The Charter provides for three Secretaries; up to the time concerned — and for many years thereafter — these were the two through whom the Council conducted most of the non-financial affairs of the Society and the Foreign Secretary who on behalf of the Council conducted formal dealings with foreign astronomers. He informed any such who was to receive recognition by the Society, he attended to the proper reception of distinguished foreign visitors and he sometimes composed congratulatory addresses to foreign academies and the like on occasions of special celebration. For this reason it was appropriate that the Society should elect an internationally known senior British astronomer, normally a past President.

'Geophysical Secretary'

1927–31 W.M.H. Greaves.

1931–45 R. Stoneley FRS (1894–1976) Lecturer in Mathematics, Leeds to 1934, Cambridge from 1934.

At this time the secretary of the Geophysical Committee was a member of the Council who was required to carry out for the Society's geophysical activities duties like those performed on the astronomical side by 'the Secretaries'. He was generally referred to as the 'Geophysical Secretary'. However, so long as the three secretaryships provided by the Charter were allocated as described above, he could not be a legitimate Secretary of the Society. Over a great many years the

Council sought ways of overcoming or circumventing this situation. At the Council meeting of 1935 December (1935 Minute 177) the Secretaries proposed, 'That the President shall discharge the duties hitherto discharged by a Foreign Secretary and that a Secretary be nominated whose duties shall be to edit the *Geophysical Supplement*, to discharge generally such duties with respect to the *Supplement*, and papers submitted for publication therein, as are performed by the other Secretaries with respect to the *Monthly Notices*, and to arrange Geophysical Meetings.'

At the next meeting of the Council (1936 Minute 16), 'After discussion, the motion in its original form was withdrawn, and it was decided to alter the Council's regulations relating to the nomination of officers and council so as to retain the secretary of the Geophysical Committee on the Council.' The outcome was that so far as election and tenure were concerned he was to be treated in the same way as the other Secretaries, but he still did not have the status of a Secretary under the Charter. It is worth mentioning that this remained the situation until 1964 when the Council replaced the Foreign Secretary by the Geophysical Secretary (although not distinguished as such from the other Secretaries on the voting paper), and appointed a member of the Council as 'Foreign Correspondent', who then did not have the status of Secretary. In due course this makeshift office was allowed to lapse.

The Society's Staff

Assistant Secretary Miss Kathleen Williams LRAM had been Clerk 1920–23 (salary £130) and Assistant Secretary since 1923 (salary £250 in 1929, £270 in 1939, with rent-free accommodation on the top floor in Burlington House).

Clerk Miss Edna Wadsworth (MBE) MA had been appointed in 1923 (salary £175 in 1929, £190 in 1939). Besides all she did in the office, she also undertook most of the routine work in the library. For trips between the office on the ground floor and the library on the first floor she must surely have held every possible record — fastest time, most trips in a day, greatest total number!

Porter and Caretaker Mr and Mrs R.J. Steel had been appointed in 1925 (wages £2.5s.0d. full-time and £1 part-time in 1925 with deduction of 12s.6d. per week for living accommodation in the basement, but they seem to have had free light and heat; also there were occasional extras for services to other societies in Burlington House on special occasions; Mr Steel had a service pension as well).

Ordinary Meetings: character and conduct

The Society held an Ordinary Meeting on the second Friday of every month from November to June (unless this happened to be Good Friday, when the meeting took place on the preceding Wednesday). Tea, coffee, bread and butter and excellent fruit cake were provided in the library from 4.30 pm. The available space became very congested, but Fellows appeared to enjoy exchanging greetings at extremely close quarters. The porter rang a hand-bell at 4.55 pm when Fellows began to pack into the Meeting Room. There were some dozen benches, each seating some eight Fellows with moderate comfort, and one or two more than that could sometimes be accommodated without comfort. The usual attendance was about 100, but up to about 130 on occasion managed somehow to get themselves into a meeting.

Until 1939 March the Society paid the Post Office about £5 per year for a time-signal which produced a single sharp 'ping' on each hour in the Meeting Room — the myth being that the signal came on a direct line from Greenwich. The Society employed Messrs Dent of Pall Mall to look after the installation. A President who failed to start the meeting by rising in his place precisely on the five o'clock 'ping' was regarded as having spoiled his copybook. His first action was to ask the Secretary to read the minutes of the previous meeting. These were extremely terse; reading them today gives no impression whatever of what a meeting was like. To gain some such impression without having been there one must turn to the pages of *The Observatory*. Any special announcement such as that of the award of the Gold Medal was then made by the President himself. With the help of the 'junior' Secretary he then conducted the formal business, each item being called for in time-honoured words; typically:

President: 'I call upon the Secretary to read the list of presents received since the last meeting.'

Secretary: '. . . presents have been received since the last meeting, including. . .

President: 'I ask you to return your thanks to the respective donors.' (While I understood that a Fellow would present his latest book, for years I pictured the other 'presents' that arrived in such quantities as flowers, chocolates and the occasional pheasant from the Society's admirers. But then there was the occasion when the Secretary's reply was, 'Fifty-seven presidents [*sic*] have been received since the last meeting.' Ultimately I was sadly disillusioned by learning that in general 'present' meant an issue of a journal sent in exchange for one from the Society.)

President: 'I call upon the Secretary to read the list of communications received since the last meeting.'

Secretary: 'The following papers have been received (authors and subjects)', or, 'Papers E547 to E565 have been received since the last meeting.' (The office kept a 'papers book' in which particulars of papers were entered in the order in which they were received, E547 being number 547 in the current book 'E' in the sequence.)

President: 'I call upon the Secretary to read the list of candidates for suspension.'

Secretary: 'Certificates have been received in favour of the following (names and addresses of candidates for Fellowship, with names of proposers and supporters).' (The list would have been to the Council the same day; 'suspension' meant display in the Library until these candidates came up for election two months later.)

President: 'I call upon the Secretary to read the names of Fellows elected by the Council subject to the confirmation of this meeting', (after the Secretary had done so) 'If any Fellow desires a ballot, now is the time to notify the fact', (after a brief pause) 'No Fellow having demanded a ballot, I declare the election of these Fellows by the Council confirmed.' (Actually there was one occasion in the decade when a Fellow did demand a ballot, but he had declared his intention beforehand. There had to be a separate ballot box for each candidate on the list, and the Fellows present were expected to vote for or against each one. On that particular occasion everyone on the list was in fact elected; so no one knew (at any rate officially) which candidate it was to whom some Fellow was objecting.)

If there happened to be present an Associate who had not yet been formally admitted to the Society, it was a duty of the Foreign Secretary to apprise the other officers of the fact. The President then called him forward, the senior Secretary directed him to sign the roll and presented him to the President who took his hand and pronounced, 'Dr . . ., in the name of the Royal Astronomical Society, I admit you an Associate thereof.' After that, 'Any Fellow present who has paid his admission fee and first contribution and has not yet been formally admitted' was invited to come forward and sign the book; the President then admitted him, using a similar formula.

Visitors from overseas were not so numerous as in the age of jet aircraft. If any established astronomer from abroad was present, the Foreign Secretary was expected to ensure that he was being offered proper attention and again to let the officers know about him. It was customary for the President to say a few words of welcome and to invite him briefly to address the meeting; the visitor would usually say a little about current astronomical activity in his country.

A pleasing incident occurred at the meeting on 1931 January 9 at the beginning of which the President (Crommelin) had announced the award that day of the Gold Medal to W. de Sitter. Before the reading of the last paper on the programme he was able to say, 'A cablegram has just come from Professor de Sitter in Leiden saying, "Please convey to the Society my best thanks and high appreciation."'

Except in the case of the February meeting the 'meeting card', posted a day or two before to those Fellows who had put themselves on the list for receiving it, consisted in general simply of the titles and authors of papers received since the previous such card (and from 1937 November it mentioned also any papers from that card that had not been 'read' at the previous meeting). If he was present and wished so to do, an author of a paper on the list could usually rely upon being allowed to describe his paper. He was expected to take no more than 5 to 10 minutes and to concentrate on the central ideas of the work. The President liked there to be time for discussion; if the work of any astronomer in the audience was in any way involved, the President would probably invite him to comment, and then anyone else could participate. It was quite usual for a senior astronomer to preface his comment or question by a word of congratulation to the author. If it looked as though a paper — particularly one by a young author — might go without remark, nearly always some senior Fellow would offer some relevant comment which might then stimulate an interesting short discussion. The Astronomer Royal, Sir Frank Dyson, was particularly good in this way, simply as one scientist to another, with no appearance whatever of being patronizing. If an author was unable to be present, he might arrange with the Secretaries for someone else to describe his paper. Or, if the Secretaries knew that the author of a paper in the list, which happened to be of topical interest, would be unable to present it, they might send it to some appropriate Fellow before the meeting and ask him to speak about it. Also the Secretaries between them might describe two or three of the papers themselves, either because they happened to have an interest in the work, or simply because they were short of other speakers. Actually such reading was a duty enjoined upon the Secretaries by a then bye-law, but as years went by this was honoured in the observance less and less frequently. In his early days of attending meetings, the present writer developed intense admiration for the then Secretaries W.M. Smart and W.M.H. Greaves for their evident mastery of every branch of astronomy with which they had to cope in this way. He is still convinced that they possessed exceptional talent, but also that a little practice had given them the knack of readily extracting from a paper the essentials of what the author had done — or claimed to have done. It should be said too that most of the leading astronomers of the time — outstandingly Eddington and Milne — were brilliant at summarizing their work on these occasions. Affixed to one side of the blackboard — and this was some 30 years before 'view-graphs' came into use — there was a black flap upon which the junior secretary chalked up the list of speakers. Those with a good reputation were given some priority, young speakers usually came next and at the end one of the senior Fellows who could never bring themselves to stop. For in those days tradition required that on the 'ping' of seven o'clock the President should declare the meeting 'adjourned' until the date of the next. Tradition took precedence over consideration for the speaker and he was left without an audience! At a fairly normal meeting of this sort as many as 10 papers could be 'read' and discussed.

Such then was the staple diet of the meetings. It sprang from the original notion of a scientific society as a body of persons with a common interest meeting together for the exchange of information and opinions. In order to give the meeting some form and order it obviously became necessary to have main contributions in the form of 'papers' that were 'read' at the meeting. For various evident reasons it was valuable to have such papers circulated to members who had been unable to attend, and maybe published more widely, but the concept of the published papers emanating from the Society's meeting was pre-served by calling the publication by some such name as *Proceedings* or *Journal.* Only after World War II did the original concept of proceedings as concerning publication, and with it the concept of a society's meeting as a presentation of the papers that composed the proceedings, gradual-ly disappear — although the war itself had brought about at any rate a temporary cessation of the old-time style of meeting. Anyhow, it is because of this change that it has seemed worthwhile here to give some picture of what things were like before the change.

Although what has been described was the regular fare, there were titbits and special treats from time to time. The George Darwin Lecture had been inaugurated in 1927 and had become a well-established annual event — or movable feast. From time to time too a meeting was devoted wholly or partly to the discussion of something of particular topical interest in astronomy, and occasionally such a meeting was held jointly with some other society. Instances of all forms of activity are noted in the next section.

Ordinary Meetings: scientific proceedings

Since not many more than one hundred persons could be crammed into the Meeting Room, since a high proportion of those present at any meet-ing attended most meetings, and since there were usually a dozen or so visitors, the majority of Fellows can have attended a meeting only very occasionally. Nevertheless the monthly Ordinary Meetings of the Society have always been the heart of its existence. This seems to be true of this Society more than of almost any other. When the meeting was devoted primarily to the 'reading' of papers — as was the case throughout this decade — this had a healthy influence upon the quality of the work presented to the Society. For it was a tremendous challenge to a young scientist to have to stand up and to expose his ideas to the judgement of nearly all the leading astronomers in the land — he took good care that they would pass muster. It was undoubtedly an ambition of nearly every astronomer in the world to attend a meeting, to see the Society in action, and if possible to participate in some way. The fame and repute of the meetings have long been maintained by the style in which they are reported in *The Observatory*; obviously the reader will find particulars there for which there is no space in the present account.

Any account of the Society at any period must, however, tell something of what went on at its Ordinary Meetings. As regards speakers and topics there were, naturally, numerous selection effects. Obviously, the speakers were mostly British astronomers describing the British astronomy of the day. However that was not all of what these astronomers were interested in. As we shall see, much of the work of some cosmologists did not come to the Society; for instance, at most meetings Eddington had some significant intervention on a plainly astronomical topic, but, apart from one or two special occasions to be mentioned, there was scarcely a clue to the fact that his main preoccupation in those years was with his 'fundamental theory', as it came to be called. Also geophysicists like Chapman, Jeffreys and Stoneley were much in evidence at the meetings, but their own scientific contributions were mostly dealt with at Geophysical Discussions. Again a feature that contributed to the character of the meetings was that a number of prominent participants were amateurs, in the sense that they worked independently with their own instruments; although some took a very active part in the proceedings, they presented much of their own work at meetings of the British Astronomical Association (BAA). Incidentally, in those times Geophysical Discussions and meetings of the BAA were quite fully reported in *The Observatory*. Lest it should then appear that the Society's scientific proceedings might tend to be unduly insular, it has to be pointed out that it was continually receiving visitors from all over the world — again probably to a relatively greater extent than other societies. In the years concerned, the George Darwin Lecturers were all from overseas; each gave a survey of an important field of work from some standpoint that gave fresh enlightenment to his audience. Besides these lecturers, British astronomers working abroad, and astronomers from abroad working here, overseas visitors who took part in Ordinary Meetings in the period included W.S. Adams (Pasadena), G. Alter (Prague), B. Bok (Harvard), P. ten Bruggencate (Göttingen), Count F.M. da Costa Lobo (Coimbra), H.D. Curtis (Michigan), T. Dunham (Pasadena), E. Hertzsprung (Leiden), E.P. Hubble (Pasadena), W.J. Luyten (Minnesota), C.E.K. Mees (Rochester, NY), S.A. Mitchell (Virginia), J.H. Oort (Leiden), J.S. Plaskett (Victoria, BC), M.N. Saha (Allahabad), H. Zanstra (Amsterdam), F. Zernike (Groningen).

As regards any frequent 'home' participant, there was naturally some selection beyond scientific attainment. He had to be able to present the essentials of a paper clearly and audibly in no more than about 10 minutes and then to give brisk answers to questions about it. In such matters tradition and example counted for much, but each speaker who qualified had his individual style: Eddington, so hesitant in conversation and dull in academic lectures, was excellent in a measured but witty fashion; Milne, on his day, was fast, fluent and boyish; Jeans and Spencer Jones were always polished, and so on. Plate 6 shows Jeans lecturing in the Society's old Meeting Room under the Presidency of Reynolds. There were one or two well-known Fellows who, given the chance to speak, did not know when to stop; when one of them had a paper, the Secretaries dealt with him in the fashion described earlier. It seems significant for the history of astronomy to remark that anybody who attended even about one meeting per year in the decade would almost certainly have seen and heard in action, in addition to the astronomers already named as Officers, and some of the visitors just mentioned, M.G. Adam, R.d'E. Atkinson, A. Beer, C.R. Burch, J.A. Carroll, S. Chandrasekhar, S. Chapman, L.J. Comrie, T.G. Cowling, J. Evershed, E.F. Freundlich, F.J. Hargreaves, H.R. Hulme, A. Hunter, J. Jackson, J.H. Jeans, H. Jeffreys, R.A. Lyttleton, G.C. McVittie, E.A. Milne, H.F. Newall, H.H. Plaskett, T.E.R. Phillips, R.O. Redman, R.A. Sampson, W.H. Steavenson, A.D. Thackeray, E. Gwyn Williams, R. v.d. R. Woolley.

Probably the two great discoveries in astronomy in this century have been those of *the expansion of the Universe* and of *the thermonuclear source of stellar energy*. Because of their timing we might have expected them to feature more prominently in the Society's scientific proceedings than was in fact the case. The second does belong to the decade, and it will be mentioned below. The empirical 'discovery' of the expanding Universe is usually reckoned to be denoted by the publication of Hubble's work in 1929, and it can be said to have been generally accepted before the start of the decade. But such a revolutionary concept should have sparked off many lines of research. Indeed at the meeting of 1931 March Eddington announced that Lemaître's classic paper of 1927 on the subject was to be translated and re-published in *Monthly Notices*. He proceeded to give a short account of a new paper

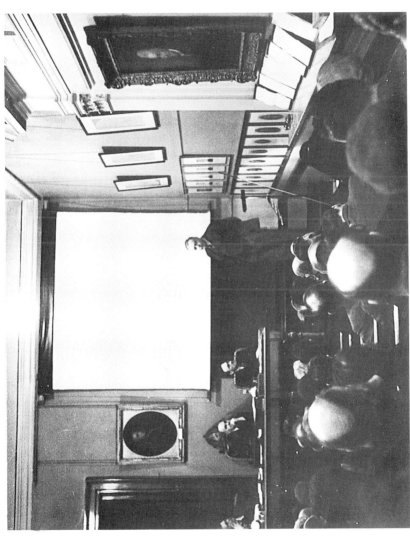

Plate 6. Sir James Jeans OM, FRS speaking at the Society's Ordinary Meeting of 1935 March 8. The Officers from the left are W.M.H. Greaves, J.H. Reynolds (President) and W.M. Smart. (Photograph by G. Merton. © Royal Astronomical Society Club.)

[*facing page* 52]

by Lemaître on the question of why the Universe is expanding and not contracting. Then at the November meeting of that year Eddington spoke about a paper of his in *Proceedings of the Royal Society* on the wave equation of the electron. He claimed that this gave a value for Einstein's cosmical constant Λ leading to a theoretical evaluation of what we now call the Hubble constant H_0. He gave $H_0 = 528$ km s^{-1} Mpc^{-1}, but called this an 'upper limit'; he said that observational values ranged from 430 to 550. There are several interesting historical features:

To begin with, I think that the awakening of interest in cosmology in Britain was not directly attributable to Hubble's work, but that it arose out of Eddington's ideas about the significance of the constant Λ. In so far as Lemaître's work had been a factor, when he wrote his 1927 paper he had been in America and his work could have been partly stimulated by what he had heard about the observations to date. Actually the fact that Eddington achieved a value of H_0 so near to Hubble's was probably a factor in causing other astronomers to accept Hubble's value. As N.B. Slater showed much later, the irony was that, had Eddington done his own calculations self-consistently, he would have obtained about Baade's empirical value of H_0 of the year 1952. As it was, at a later meeting in 1935 December Eddington gave 136×2^{256} as the total number of protons/electrons in the Universe. He went on to reaffirm his conviction that the value of H_0 could be settled by theory, declaring 'when it has been settled I shall know whether to believe the observational value'.

The legend of rivalry between Eddington and Jeans is well-known; occasionally they mischievously fostered it. But when Eddington produced the ideas just mentioned, as he readily acknowledged, Jeans was generous in complimenting him at the meetings. In fact, nobody ever understood this part of Eddington's work!

During the decade rather few 'papers' on cosmology were read. On the observational side this was obviously because effectively all the work was carried out in America. On the theoretical side much of the work was done in Britain but there was a tendency to regard it as somewhat too mathematical for *Monthly Notices* and quite a lot went elsewhere. However the subject kept claiming attention at meetings. At the Annual General Meeting in 1933 Knox-Shaw gave the first Presidential Address of modern times not devoted to the work of the year's Gold Medallist; he spoke on 'The distances and motions of the extragalactic nebulae.' At the close of an able review of the empirical situation he declared 'the hope that British astronomy, to which we owe so much of the pioneer work on the nebulae... may someday be able once again to play its part in contributing to this most fascinating branch of our science.' Although he did not mention it, Knox-Shaw was busy planning for the erection of the new Radcliffe telescope in South

Africa. Then at the meeting of 1934 May there was a discussion on 'The Expanding Universe'. Milne opened it with a brief account of 'Newtonian cosmology' recently developed by him and McCrea; there followed contributions from Hubble, Lemaître, Shapley, Oort and McCrea. Under the same title another discussion was held on 1939 January 27 in the Royal College of Science at a joint meeting of the Society and the Physical Society. The President of the Physical Society invited the Astronomer Royal (Spencer Jones) to conduct the proceedings: speakers were G. Temple, McVittie, Jeans, Eddington, McCrea, W. Wilson. Deliberately or not, the discussion constituted a review of the situation — chiefly on the theoretical side — 10 years after Hubble's initial publication. The account in *The Observatory* **62**, 67–73, 1939 forms an interesting record of the state of cosmology in which it remained effectively frozen until after the war.

The whole field of *the structure of stellar interiors and stellar atmospheres* was another in which a carry-over from the preceding decade seemed inevitable. The dramatic developments in understanding the basic principles had been argued out at the Society's meetings more than anywhere else in the world. It looked as though the saga might carry on indefinitely when the first meeting of the new decade in 1931 January was devoted almost entirely to the postponed discussion of a paper Milne had read in the preceding November. That had presented probably the most comprehensive account to date of Milne's criticisms of the work of others — chiefly, of course, of Eddington — and of Milne's own proposed stellar models. Nearly all the chief participants in the earlier debates were present and went into action. Eddington and Jeans were outspoken in opposition to Milne. Cowling had a more constructive comment, while F.A. Lindemann (later Lord Cherwell) came out in favour of Milne. The great pure mathematican G.H. Hardy declared 'a mild preference for Milne's theory', but for a purely mathematical reason. Stratton and Greaves had interesting comments. The physicist Sir Oliver Lodge preferred to concentrate upon what was common ground for all the astrophysicists and upon the progress denoted thereby. As it turned out, this was the last debate on such a level about stellar structure, and the last meeting of the Society to bring together all these particular individuals. Through the new decade many important developments in the theory of the internal structure of stars continued to be first announced at the Society's meetings but they sprang from physically more realistic models constructed with particular applications in view, rather than from any over-ambitious, all-embracing, scheme. Thus, for instance, Milne, Chandrasekhar, Cowling and others presented results on composite stellar models, distorted models, stability of models. Eddington, Cowling and others discussed the role of convection and the crucial question of hydrogen abundance in stars. In this last connection, it is interesting that in 1939 June one conclusion in a paper by Eddington was 'hydrogen is rather more abundant than usual in both Sirius and its companion, with perhaps a slight excess in the companion'! (author's exclamation).

The most notable debate in the decade concerned *relativistic degeneracy*. In 1926 R.H. Fowler had accounted for the nature of white dwarf stars in terms of the equation of state of a degenerate (electron) gas. A few years later physicists, notably E.C. Stoner, had claimed to show how this equation has to be modified when relativistic effects are taken into account. S. Chandrasekhar then studied the implications of the new equation of state for the theory of stellar structure, and in particular for the theory of white dwarfs. At the meeting of 1935 January he presented his now classic paper in which, amongst other things, he propounded the existence, properties and significance of the 'Chandrasekhar limiting mass'. The very next paper was by Eddington on 'Relativistic degeneracy.' 'I do not know', he began, 'whether I shall escape from this meeting alive, but the point of my paper is that there is no such thing as relativistic degeneracy!' That meeting has become an unhappy legend. Although both papers had been announced on the meeting card, and although the authors were in daily contact in Cambridge, Eddington had deliberately refrained from revealing what he was going to say. I am convinced that he believed it to be fairer and kinder to his young colleague to let him present his ideas unhampered by knowledge of what was to come; he had even gone to the length of arranging with the officers for Chandrasekhar to have a whole half-hour in which to do so. But others hold that Eddington was wrong to keep him in the dark about his intentions. Everyone now seems agreed that Eddington was mistaken as judged by still standard relativity and quantum theory, but whether that theory is the last word is a different matter. Anyhow he persisted in his opinion and reaffirmed it in his paper of 1939 June already quoted. It is alleged that Eddington's stand delayed progress in the field for about 30 years. The present review of the history of those times makes me think that other factors were mainly responsible.

One striking circumstance is that at none of the meetings then did anyone seriously tackle the problem as to what *does* ultimately become of a star of mass exceeding the Chandrasekhar limit. At the meeting of 1936 May Chandrasekhar himself mentioned some speculations on the subject, but only rather incidentally; otherwise I have not noticed it ever being raised in connection with the debate. The truth seems to be that astronomers at the time were not seriously evolution-minded. This was not for lack of interest but for lack of knowledge upon three basic factors which would enable some of the right questions to be asked about evolution. These were regarding the hydrogen content of the stars, the mechanism of stellar energy generation and the time-scale of the Universe. We have seen that the first of these did come before the Society but that knowledge remained uncertain up to the end of the decade. The third was much in the minds of Fellows; for instance, a large part of the meeting of 1935 March was given to a discussion on 'The age of the Universe', with Jeans, Eddington and Milne as the speakers. As regards the second, the problem was frequently referred to but, to my knowledge, it was never the main topic of a paper or discussion in the decade. This was because the basic discovery of the crucial nuclear processes when it was announced in 1938/9 by C.F. von Weizsäcker and by H.A. Bethe, was not given in any communication to the Society; also it was too late to have much impact in what remained of the decade before the outbreak of war. (The Society did publish an account in *Occasional Notes* **1**, 78−88, 1939.)

In the rest of *solar and stellar astrophysics* much of the observational work was ever more refined *photometry* yielding contours of spectral lines, line-intensities, line-shifts, and gradients in continuous spectra — often expressed as colour indices or colour temperatures. Much of the theoretical work was on the structure of the outer layers of stars, on the formation of spectral lines, and on applications of the theory to the interpretation of all the new observations. Almost every meeting had papers in this field. It is difficult to pick out items for special mention, but it may be of interest to note that in 1936 May Saha remarked 'I suppose that the Lyman-alpha line will appear in the solar spectrum as an emission line', and he mentioned balloon experiments that might check such predictions. One believes, however, that this had to await observations from 'above' the atmosphere. They showed that Saha had been right! Another indication of topics of investigation is the new terms that were coming into currency, e.g. 'curve of growth', 'supergiant,' 'supernova'.

In this general line of work, from time to time the Society or an individual Fellow would unofficially adopt a particular astronomical object. The meetings would receive up-to-date reports on its behaviour. The most famous case before the war was Nova Herculis 1934. The distinguished meteor observer J.P.M. Prentice discovered it early on 1934 December 13, and he informed the Royal Observatory forthwith. Spectra were obtained there and elsewhere and shown at the Society's meeting the next day! Apart from the Sun it became undoubtedly the most observed (spectroscopically) star ever. The object itself responded with a varied and prolonged performance — at one stage observers believed it had split into two stars. Papers on it continued to be presented throughout the rest of the decade. In Cambridge it became a point of honour never to miss the chance of obtaining a spectrum; at the meeting of 1935 March, J.S. Plaskett congratulated Cambridge on having by then secured more spectra than any observatory in America. It would probably be rewarding for some astronomers to re-examine the great accumulation of observations in the light of modern theory of novae.

We shall see how the Society along with the Royal Society through their Joint Permanent Eclipse Committee was responsible for organizing British eclipse expeditions. As things turned out the total eclipse of 1936 June 19 almost certainly marked the all-time peak of such efforts. Apart from the famous 'Einstein effect' occasion of 1919, this was (one believes) the only time when the main reports by the observers were presented to a joint meeting of the two Societies — on 1937 May 27.

About a century earlier the Royal Observatory had begun to concern itself with the Sun because of its influence upon terrestrial magnetism. In the 1930s there was much renewed interest in solar–terrestrial relations, partly because of effects upon the ionosphere and partly because of the new facilities for observing solar activity. Workers at Greenwich and at the Solar Physics Observatory regularly reported progress at the meetings, and the Society also began to hear important papers arising from the operation of Plaskett's new installations at Oxford.

There were papers on all aspects of *Solar System astronomy*. Led by Spencer Jones there were comprehensive observations of positions and motions of the minor planet Eros designed to yield a new determination of solar parallax. Reports on progress of the work came to several meetings, but the final publication of the results came in the next decade. The planet that was most observed for its own physical behaviour was Jupiter. The meetings heard a good deal about this and, over the years, the Reverend T.E.R. Phillips reported the results of a lifetime's observations by A. Stanley Williams (1861–1938). This was one factor leading to the holding of a discussion on 'Planetary atmospheres' at the meeting of 1939 April; speakers were Spencer Jones, Phillips and Reynolds. There were also occasional papers on the constitution of the planets. In 1936 April Smart described a paper by R.A. Lyttleton on the 'Origin of the Solar System'; he discussed the possibility that the Sun had been a binary star and that the planets resulted from a collision suffered by the companion with an intruding star. The theory continued to attract interest at some subsequent meetings; related thereto was Lyttleton's paper on the rotations of the planets in 1938 December.

Turning to so-called *'fundamental'* astronomy, the meetings always responded to hearing a report by some noted and enthusiastic professional about the successful culmination of some professional observing programme. They had major reports on parallaxes, proper motions and related matters, chiefly from Greenwich, the Cape of Good Hope and the Radcliffe Observatory. They heard papers on the interpretations of such work in terms of galactic rotation with, in particular, some good — and some wildish — evaluations of the Sun's motion about the galactic centre. With an important bearing upon such matters there was work on *interstellar matter*. There appears to be a widespread impression that such work began in earnest only after World War II; in fact in the 1930s much work was reported on interstellar absorption lines, space reddening of stars and the inferred distribution of the material causing it. Indeed in 1938 May Charles Fabry of Paris entitled his George Darwin Lecture 'Interstellar space'; speaking as a physicist, he gave a comprehensive review with a still modern ring about it.

The Society has always held instrument-makers in high esteem. It has not been a forum for technological aspects, but it has welcomed papers on developments that exploit some new idea, or report new exploitations of known techniques. Surprisingly many things ever since treated as routine were mentioned as novelties at meetings in the 1930s. Thus there were papers on new methods of testing (particularly convex) mirrors, on liquid prisms, on photoelectric photometry, on the uses of the spectrohelioscope, on the new Oxford solar telescope, and visitors from Pasadena reported progress on the 200-inch telescope. B. Lyot gave the George Darwin Lecture of 1939 on 'The Study of the solar corona and prominences without eclipses.' Lyot showed impressive cinematograph films of solar activity, and in the previous two years the Society had seen pioneering films of this sort made at the McMath – Hulbert Observatory, Michigan. The use of colour photography in astronomy was mentioned more than once, but no British results were described. The first use in Britain of aluminium, instead of silver, as a coating for telescopic mirrors was reported in 1934. There were papers on clocks and the use of quartz clocks in astronomy had its first mention. At the meeting of 1939 January H. Zanstra described probably the first experiment in 'laboratory astrophysics'. At the meeting of 1932 April L.J. Comrie recalled that the Society had awarded its first Gold Medal to Charles Babbage for his projected calculating machine, and he went on to demonstrate the Burrough's machine that had been in use for a few years past in HM Nautical Almanac Office of which he was Superintendent — apparently the first use of any such machine in a scientific establishment in Britain.

Finally, brief descriptions of a few specimen meetings show how things actually happened.

1934 May, President: Stratton. After formal business Harlow Shapley delivered the George Darwin Lecture 'Some structural features of the metagalaxy'. The President gave an address on the award of the Gold Medal (quoting a passage from Lucretius) and then presented it to

Shapley. On Stratton's invitation, Edwin Hubble, representing the Astronomical Society of the Pacific, occupied the Chair, gave an address and presented that Society's Bruce Gold Medal to Alfred Fowler. Stratton resumed the Chair and conducted the discussion on 'The expanding Universe' already mentioned as having the participants Milne, Hubble, Lemaître, Shapley, Oort and McCrea. How so many overseas astronomers came to be there is difficult to discover now, but the Society certainly took good advantage of their presence.

1937 April, President: Spencer Jones. The president announced the election as Associates of Lyot, Oort, Shajn; he gave notice of the joint meeting with the Royal Society to be held in May. Woolley described a paper by himself and L.S.T. Symms on 'Two visual binary orbits'; four Fellows commented. A paper by H.P. Robertson announcing what has since become known as the 'Poynting–Robertson effect' was described by McCrea; there were twelve questions/answers. H. Horrocks spoke on the theory of the figures of non-uniformly rotating polytropic stars. Redman described photometric work by E.G. Shirley and himself on the 'nebulae' M31, M33 — work of importance in connection with Hubble's distance-scale; there were two comments. Milne returned to polytropic stellar models and spoke at length on the problem of fitting photospheres to them. R.T. Gunther, Director of the Oxford Museum for the History of Science, gave a talk on historical astronomical instruments recently discovered at Oxford.

1939 November, President: Plummer. The first part was a Special General Meeting at which emergency bye-laws were proposed, discussed and adopted. The rest was the Ordinary Meeting. Commemorative resolutions on the recent deaths of two very senior Fellows, Crommelin and Hollis, were read. G.L. Camm described his paper on 'The Sun's speed of galactic rotation determined from the globular clusters'; Eddington was 'a little sceptical'. H.R. Hulme described a paper by D.R. Bates, 'The quantal theory of the continuous absorption of radiation by various atoms', probably the first paper to come to the Society that depended on computations using Massey's 'differential analyser' at University College. Hulme also called attention to 'a very important paper by R. Wildt' (not before the Society); Massey had calculated the absorption coefficient of the negative hydrogen ion; Wildt showed how this could account well for the form of the continuous spectrum of stars like the Sun. Thackeray described C.W. Allen's papers 'Stark effect and damping factor in the Fraunhofer spectrum' and 'Photometry of the solar D lines'; several Fellows joined in a lively discussion. D.R. Barber gave his paper 'A sensitometer study of some developers and emulsions of astrophysical interest.' Hulme read a paper of his own, 'The law of error and the combination of observations', the President remarking that he had just published a book on the subject and he thought Hulme would find it in agreement with his paper.

These examples are intended to illustrate the varied entertainment to be enjoyed at a meeting, and to give some indication of trends in British astronomy when the war was starting. So much variety was fitted into the bare two hours because no time was wasted. In general a speaker used no aid other than either blackboard and chalk or projection of astronomical photographs — there were no paraphernalia of microphone and viewgraphs! The scope of astronomy expands and the style of presentation of its results has to change. But the same spirit of eager anticipation that informed these meetings of long ago ought never to be lost.

Anniversary Meeting

The meeting on the second Friday in February was the Society's Anniversary Meeting. The first part counted as an Ordinary Meeting. It began with the same sort of business as any other such meeting, but for the rest it was given over entirely to the President's Address, or Addresses, and the presentation of awards.

The Society's Gold Medal was regarded as the world's highest award for astronomy and the Council has always taken immense care in selecting the recipient for the year; it relied upon its President publicly to state clearly the grounds for the selection. So the practice had grown up of the President devoting his entire address to an account of the work of that year's medallist. However, when J.H. Jeans endowed the annual George Darwin Lecture — the first of which was given in 1927 — he suggested that if the medallist was resident overseas, it might be appropriate to invite him to deliver that year's lecture. In that case he would be expected himself to give an account of some of his work, and to do so with more expert knowledge of it than even the President of the Royal Astronomical Society could claim. Under that arrangement, the President need give only a short address on the award, either at the Anniversary Meeting or else on the occasion of the George Darwin Lecture, and devote his main address to a topic of his own choosing. Also it had previously been the practice for the President at the Anniversary Meeting to hand the medal to the medallist or his representative (usually someone from his country's embassy) but under the new scheme a medallist could more often receive it in person from the hands of the President. Since it was envisaged that the George Darwin Lecturer would normally be someone from overseas, it was at first supposed that the traditional procedure would still be followed when the medallist was from this country. In any case there could be no rigid rule since a medallist from abroad could not be expected always to be able to give a lecture to the Society in the year of the award. Besides all this, traditions die hard and there were Presidents and others who preferred the old practices.

Sir Arthur Eddington was one who tried to hasten the demise of these traditions. At the Council meeting of 1932 May 13 (1932 Minute 71) he presented a memorandum 'On Presidential Addresses and the Annual Meetings' which is surely a model document of this sort. One would like to quote it in full, but to be brief one can say simply that it sought to give more recognition to the newer practices just mentioned with the particular object of ensuring that a President should give one (full-length) Presidential Address only in his term of office, and that most of the time of the Anniversary Meeting should be given to reading of papers or to a discussion. At the next meeting (1932 Minute 93) 'the procedure outlined in the memorandum was approved,' and the President (Knox-Shaw) announced it at the 1933 Anniversary Meeting. Knox-Shaw proceeded to give a 'new-style' Presidential Address at that meeting and Stratton who then succeeded him gave 'new-style' short addresses on the award of the Gold Medal immediately after the George Darwin Lectures delivered by the recipients in that year and the next. However, Stratton in 1935 reverted to the 'old-style' of address when the Gold Medal went to E.A. Milne who, as a resident in the United Kingdom, was in those days regarded as ineligible to give the George Darwin Lecture. Since then every President has felt required to give two addresses as previously, even though, as stated, the intention of Eddington's memorandum in 1932 had been different. So far as the decade under review was concerned what actually happened as regards Gold Medallists and George Darwin Lectures is summarized in Table 1.

In 1940 the Gold Medal went to E.P. Hubble of California and, on account of the war, there was that year no George Darwin Lecture by him or anyone else (Hubble gave the lecture in 1952). So again the President, this time H.C. Plummer, delivered an 'old-style' address on Hubble's work on 'the distances, velocities, distribution and nature of the extra-galactic nebulae'. At the close he mentioned, 'Of the 48 medals which have been awarded in the preceding 50 years 18 have been given to British subjects, 16 to the citizens of the United States and 14 to other nationals. But America has a claim to...Ernest Brown...If it be granted, it can be said that over the half-century the balance between the two branches of the English-speaking race has been exactly even. In the last ten years this is the fifth medal to go to America.'

In a year in which the Jackson Gwilt Medal and Gift was awarded the President spoke briefly about it and made the presentation. Some Fellow who had been so commissioned by the Secretaries then thanked the President for his address(es). This part of the proceedings took until about 6 pm, when the President requested visitors to withdraw so that

Table 1.

| President | Address | Medallist | Presentation | George Darwin Lecturer |
| | M: on the award of Gold Medal | | A: Anniversary | M: Medallist |
			D: George Darwin Lecture	
1931 Crommelin	M	W. de Sitter	A(r)	M
1932 Knox-Shaw	M	R.G. Aitken	D	M
1933 Knox-Shaw	Extra-galactic nebulae	W.M. Slipher	D, Stratton	M
1934 Stratton	International cooperation	H. Shapley	D, Stratton	M
1935 Stratton	M	E.A. Milne	A	H.N. Russell
1936 Reynolds	Galactic nebulae	H. Kimura	A(r), Reynolds	A. Kopff
1937 Reynolds	M	H. Jeffreys	A	N.E. Nörlund
1938 Spencer Jones	M	W.H. Wright	A(r)	C. Fabry
1939 Spencer Jones	Meridian astronomy	B. Lyot	D, Plummer	M
1940 Plummer	M	E.P. Hubble	A(r)	None

In the Presentation column the name is that of the President giving a short address on the award
(r) denotes that the medal was handed to the medallist's representative

the Society could proceed to the business of the Annual General Meeting, of which all Fellows had received due notice.

After agreeing to the minutes of the last Annual General Meeting (and any intervening Special General Meetings), if notice had been given of any proposed change of bye-laws, the Secretaries on behalf of the Council moved their adoption. Usually the Council's proposals were adopted. Then the auditors of the Treasurer's accounts for the past year, who had been appointed by the Ordinary Meeting in December, presented the Treasurer's accounts and their own report on the state of the Society and of its premises and property. The Treasurer himself might supply a few explanations but he was not expected to make a statement at the meeting. The auditors were simply two knowledgeable and public-spirited Fellows, who carried out their onerous task over several days each January, and they thoroughly merited the vote of thanks then accorded to them. The tradition of this vote has been preserved ever since in the agenda of the Annual General Meeting and the 'Honorary Auditors', as they became in 1940, have continued to deserve it, even though they have had little to do with the actual finances since that year. In 1939 December a firm of professional accountants was employed for the first time — at a fee of five guineas — although the accounts for 1941 were the first to be signed by them.

In those days there was not a 'narrative' report of the Council, but at some stage in the meeting one of the Secretaries would describe in a sentence or two the contents of what was then called the Report of the Council. Again having been commissioned by the Secretaries, two Fellows after being called upon by the President proposed and seconded that the Report of the Council, the Treasurer's accounts and the addresses by the President should be printed and published in the usual manner, this being invariably agreed with acclamation. All this material, together with Obituary notices, Proceedings of Observatories, Reports on the progress of astronomy, information about the Library, and so on, became the February issue of *Monthly Notices* which came out only six or eight weeks later.

The President then appointed Scrutineers, and the Society proceeded to the ballot for the new Council. Since in those days the printed list never contained any but the existing Council's own proposals for the full number of places, this was essentially only a formality. A Fellow could erase any name and substitute that of any other eligible Fellow, but in practice this was never done with any effect upon the outcome. Anyhow the business was carried through with due solemnity until the President sought leave to declare the ballot closed. While the voting papers were being scrutinized, some mild diversion was provided. One or two representatives of the appropriate committees might exhibit recent additions to the Society's collections of slides or instruments, or a Fellow might impart some astronomical news, e.g. in 1931 F.J.M. Stratton spoke of 'M. Lyot's success in detecting the corona without an eclipse'.

The Scrutineers having reappeared, at the President's bidding one of them read the list of persons elected as the new Council. Thereupon any retiring officers stepped down and their successors stepped up. The first act of the re-elected or newly elected President was to call upon two Fellows to propose and second a vote of thanks 'to the President/ Vice-President(s)/...and members of the Council who now retire'. Thanks were then given to the Scrutineers, and the meeting was adjorned until the next anniversary.

This takes time to recount, but in practice it was all carried through with expedition. At the time of writing, half a century later, the Annual General meeting takes much the same form but it lasts twice as long.

Council Meetings and Business

On meeting days the Council was summoned for any time between 2.15 pm and 3.15 pm according to the Secretaries' estimate of the time required in order to dispose of the agenda by tea at 4.30 pm. On the whole the timing was good but, since the President, flanked by the Secretaries, never failed to be in his place to open the Ordinary Meeting on the stroke of 5 pm, there were occasions when the Council had very little tea.

The new Council was elected each year at the end of the Anniversary Meeting and it met for the first time in March. Up to 1931 May the Assistant Secretary, as instructed by the Secretaries, wrote the minutes by hand in the Council minute book. The pages were numbered though not the minutes themselves, but the topic of each minute was clearly noted in the margin, which made reference back fairly easy. The first business of the Council was to have these read, confirmed and signed, and in March the minutes of the Annual General Meeting were read and approved, ready for signature at the next Annual General Meeting. From the meeting of 1931 June the minutes were duplicated from a typed stencil so that they could be circulated beforehand and taken as read at the meeting. The President signed one copy that was afterward gummed into the minute book. The Ormig Rotatory Reproducer cost £20 and continued in use until a photocopier was acquired in 1976. From the time the minutes were typewritten they were numbered consecutively through each calendar year, but the marginal notations were discontinued.

It must have been about the same time that there was a change too in the form of summons to the Council meeting. Hitherto the Assistant Secretary had used a printed agenda paper, in which routine items,

'Minutes of last meeting', 'Treasurer's business', etc., were already entered; there were spaces for 'Date and time of meeting', 'Special items', 'Other business'. Following instructions from the senior Secretary, she filled in the spaces by hand. It had been all nicely concise. Thenceforward, she sent out instead a full duplicated typewritten agenda paper, in which, as time went on, the Secretary asked for more and more details of the business to be included.

Normally the Treasurer's business came next and it began by having the President sign all the bills approved for payment. In March the Treasurer then reported the Fellows who were in arrears with their annual dues for the current and two preceding years — the '6-guinea defaulters'. The Council ordered the names to be suspended in the meeting room and instructed the Treasurer to send a final registered letter to each. Any who had not paid by the date notified in the letter was reported in due course, and his name was removed from the list of Fellows as required by the bye-laws. However, at subsequent meetings throughout the session the Treasurer had frequently to report cases of a Fellow making a special plea of one sort or another, e.g. he might discharge part of his indebtedness and ask for a longer time in which to pay the rest. An astonishing number of Fellows had serious trouble in affording two guineas a year, which for some could have been as much as half a week's income. The Council was as sympathetic and helpful as it could be in such cases.

The next regular item was the reading of nominations of 'candidates [i.e. for Fellowship] for suspension'. The certificates were then displayed in the Library for the ensuing two months; earlier they had been hung in the Meeting Room — hence 'suspension'. Then those who had been 'suspended' two months before came up for election by the Council subject to confirmation by that day's Ordinary Meeting.

The Secretaries then reported the 'communications [i.e. papers] received' since the previous meeting. With regard to each they might (a) recommend acceptance, (b) report that they had consulted one or more referees and announce the outcome, (c) request direction by the Council regarding referees. In cases (a) and (b) the Council would resolve whether to accept, subject maybe to certain emendations of the paper. Then the Secretaries reported upon cases under (c) from previous meetings. No properly communicated paper was rejected save on the advice of two referees. At the beginning of the decade, so far as possible referees were chosen from the Council itself but with the expansion of astronomy, astrophysics and geophysics this was rapidly becoming less feasible. It was partly on this account that in 1935, after careful consideration of several drafts, the Council brought into use the referee's form that has remained in use ever since without major change. The Royal Society and the London Mathematical Society were already using a form; E.A. Milne, who had originally proposed that the Council should follow the practice, told the writer that it had preferred the Mathematical Society's version to that of the Royal Society.

The business of papers was normally by far the most exacting part of the meeting. Since the Council minutes, as they had been signed by the President, were at the ensuing Ordinary Meeting 'laid on the table' for any Fellow to consult, it was important that they should not divulge details such as the names of referees; this was particularly so in relation to papers by members of the Council, who absented themselves from the Council's discussion of their papers. Therefore the formal wording of the minutes conveys no impression of the extremely careful discussion of the papers and reports thereon that occupied a large proportion of the time. And the amount of revision called for by referees increased in the course of those years. There were even several cases in which eventually some senior member of the Council voluntarily rewrote entire papers from overseas authors.

The other fairly regular fare consisted of reports of committees; these are treated later in this chapter in appropriate sections. Then, other than any that might have been handled by these committees, there were miscellaneous applications for exchange of publications, for replacement of publications lost by mishap, for loans of blocks for diagrams and endless other like forms of assistance. The response of the Council was invariably understanding and generous. It is worth noting as an example that according to 1938 Minute 27 it was in connection with a letter from Sir Sidney Burrard requesting some 'information about Reinmuth's object', read at the meeting of 1938 January 14 that the Secretaries first announced their proposal that 'the Society should from time to time print leaflets giving astronomical information in semi-popular form.' This led later in the same year to the first appearance of *Occasional Notes*, the forerunner of the *Quarterly Journal*.

The only persons whose tolerance may have been abused were the Society's miniscule office and library staff and the Secretaries themselves. The latter must have done much of the correspondence and, except for what was formally minuted, most of it must have gone unrecorded.

It was quite usual for a Council member to give notice of a motion he wished to propose at the next meeting; for example, at the meeting of 1931 January 9 'Sir Arthur Eddington gave notice of a motion that smoking be permitted at meetings of the Council.' At the following meeting 'It was resolved that smoking be permitted after 3.30 p.m.' This remained in force until 1938 January 14 when 'It was resolved that smoking be permitted from half an hour after the commencement of the meeting of the Council.' It should be recalled that, amongst the sort of people who served on such a body, smoking was much more prevalent than it became after World War II.

On a more serious level, some of what turned out to be major items of business were several times initiated in this way. An instance of this was indeed immediately before Eddington's motion on smoking, when E.A. Milne had a motion which led to a discussion of the procedure for 'the annual nomination of the new Council'. About a year later this resulted in the setting up of a subcommittee composed of S. Chapman, L.J. Comrie, H. Dingle and E.A. Milne to make recommendations. They presented a detailed report on 1932 May 13 and, with certain amendments, this was approved the following month. It resulted in a drastic overhaul of the procedure, with the aim of achieving a reasonably rapid and regular change in the composition of the Council; also it laid down a full voting procedure to be followed by the Council in drawing up its list of nominations. The amount of work that went into this undertaking was immense. The procedure thus adopted was followed without much change until about 1970 when steps were taken to give the general body of Fellows a more effective choice of those whom they elected to the Council.

There was naturally a calendar of special business. At the March meeting, standing committees for the year were appointed — nobody whom the Council put on to any committee seems ever to have demurred to serve. At the same meeting the Council 'took into consideration' the nomination of Associates — at that time a bye-law restricted the total number to not more than 50, and so the first thing to decide was the maximum number to be elected that year. In 1931 it was decided to elect none. The following year five were elected in accordance with procedure then in force by which the Council's selection had to be confirmed by the Ordinary Meeting in June. Then the Annual General Meeting in 1933 agreed to a change of bye-law that entrusted the election entirely to the Council. During the remainder of the decade it elected in April or May from those nominated in March a number not exceeding the agreed maximum. In the 10 years 1931−40 the total number elected was 22.

At the November, December and January meetings the award of the Gold Medal took precedence over all other business. In November candidates were proposed and seconded; in the years considered the number of names was between four and six. It was a standing order of the Council that no one should be proposed for the award of the Gold Medal while a member of the Council, unless the proposal was supported in writing by at least six members of the Council. This provision was rarely invoked, and never during the 1930s. In December, after due deliberation, one name was selected; in January this one individual was voted upon. In the decade the majority of the size required by standing orders for confirmation of the award to that individual was always forthcoming. Indeed, if the President had no doubt that the outcome at that stage was a foregone conclusion, he might ask if the Council would agree to a show of hands. Then, if clearly the requisite majority voted in favour, the award was declared to be confirmed. It was then announced at that day's Ordinary Meeting. This procedure, however, was contrary to the Council's own standing order that required a ballot for confirmation; from time to time some members objected to the show of hands, correctly maintaining that every member ought to vote without knowing how any other member was voting. Directly after the determination concerning the Gold Medal the Council took into consideration the appointment of the George Darwin Lecturer for the year.

In three years in the decade there were awards of the Jackson Gwilt Gift and Medal, which were made following the same procedure as for the Gold Medal. In 1931 the award went to C.W. Tombaugh in recognition of his discovery of the planet Pluto in the previous year.

Some members of the Councils of the 1930s and 1940s who went on to serve again in later decades have had a firm conviction that there had been in those earlier years more time for due 'deliberation' on matters like the award of the Gold Medal. Comparing durations of meetings then and later, the writer doubts this but he is ready to believe that people of an earlier generation expressed themselves more succinctly, in the time at their disposal! Anyhow the growing range of astronomy and geophysics made extended deliberation necessary and was in due course to cause the Council to appoint panels to assist in this.

In November Fellows were proposed and seconded as new members of Council, but further nominations were received by the Secretaries up to about the end of the month. The nomination of officers and Council was 'taken into consideration' in December.

At the January meeting applications for renewal of loans of instruments belonging to the Society were considered. This served as an annual roll-call of the instruments; if any borrower failed to hand back an instrument or properly to apply for renewal of the loan, the Council needed to know the reason why!

Towards the end of the minutes of the February meeting there was always one that read simply, 'The Council proceeded to the consideration of the Annual Report.' As already mentioned, in those days there was no 'narrative report' — which incidentally renders the task of the historian more difficult than it is for the years after it was introduced and became itself a year-to-year history. So the officers' presentation of the report to the Council was little more than a formality that regularized the formal presentation to the Annual General Meeting an hour or so later. In particular the 'Reports on the progress of astronomy', commonly called 'Council Notes', had mostly become a matter of tradition; the Secretaries seem, to have ensured that they were provided by suitable authors, with an occasional additional item of topical concern. Later in the period, the Council as a whole started to take more active interest. In 1937 January (1937 Minute 26) it resolved that the contents of the Annual Report should be considered each year at its June meeting, and it acted accordingly in 1937, 1938 and 1939; in 1938 it decided henceforth to include a 'Note' on a geophysical topic. In presenting the report of the Council for the year 1937 at the Annual General Meeting in 1938 the Secretary explained the changes. He named the special subjects dealt with in that report as: extragalactic nebulae (Reynolds), interstellar absorption (Rosseland), line profiles (Redman).

In those times the Society's rooms, as well as those of all the other societies in Burlington House, were open on Saturdays up to about 1 pm. On the Saturday after a meeting day the Secretaries came in early and, together with the Assistant Secretary (who did not attend Council meetings) cleared up all the immediate business arising out of the previous day's deliberations — a truly formidable morning's work.

The 'recess' was from June to November. At its June meeting the Council empowered the officers to act for it in well-specified ways during that time. The Treasurer paid bills approved by the President; the Secretaries saw to the refereeing of papers and to the publication of any approved by referees; such papers came out in the supplementary issue of *Monthly Notices*, published about the end of October.

Council: committees

The 1930s turned out to be a time of intense activity of the Society on its domestic side — probably unparalleled by any other time until the late 1970s. Why this should have been so is difficult to explain. It was very far from being a time of affluence. Most certainly it was not a time of leisure from other pursuits; it has already been remarked how all those involved were exceptionally busy in their scientific and professional careers. Also one can point to no single individual who bestirred all the rest, although one is bound to note how most of the undertakings were carried through successfully with the wise support and guidance of J.H. Reynolds who figures so continually in the narrative. There is no evidence, however, that Reynolds or any other one Fellow was a chief instigator of action. Rather it was a time when the Society was singularly fortunate in having in its councils a group of some score of exceedingly devoted Fellows who were inspired with the ideal of putting all its affairs into the best possible order, and who happened to collaborate exceptionally effectively.

The production of publications, and of slides and prints, the working of the library, the disposal of instruments, the maintenance and decoration of the premises, the insurance of their contents, besides various aspects of the business of the Council and the procedure operated to ensure its own continuance, all came in for vigorous overhaul — nearly always beneficial.

Looking back it is as though the Society was impelled quite literally to put its house in order before the cataclysm. As one who six years later was concerned in the effort of as soon as possible getting the Society back to normal working, the writer can now appreciate how tremendously more difficult the task would have proved had not nearly all the Society's affairs and possessions been brought to such excellent order by the time the war started.

The story of how all this came about is best told as the story of the workings of committees of the Council. These put life into the somewhat routine proceedings described in the preceding section. As stated, the committees were appointed annually in March; the Fellows selected did not need to be members of the Council for the time-being, and there was no rule about length of service. There were four 'business' committees: Library, Photographic, Finance, House. The two Secretaries were appointed to two each, the Treasurer to all four.

During the years concerned the Council set up also at least nine committees to advise it upon special business. On average, each had about a half-dozen members, it dealt with its assignment in one or two meetings, and it reported very expeditiously. One does not know whether there had to be much correspondence between members besides the actual meetings, but this would cause no undue delays in those times when letters were commonly delivered on the day of posting. One or two of these special committees have already been mentioned. Some account of the activities of the others has its natural place in the following sections.

Here we may mention one that came near the end of the period. On 1939 April 14 the Council set up a committee to consider the institution of an annual Christmas lecture for children. It was asked also to consider any suitable celebration of the tercentenary of the first observation of a transit of Venus across the Sun's disc. At the meeting of 1939 June 9 the Council adopted the committee's recommendation for a lecture to be given near Christmas 1939 by the Reverend T.E.R. Phillips (President RAS 1927-29, Jackson Gwilt Medal 1918), and for any decision about an annual lecture to be deferred until after that. The Council's resolution to hold a soirée once in each President's term of office also arose from the report. Sadly, of course, the outbreak of war caused the Council to decide on 1939 October 13 that under the conditions of that time no Christmas lectures or soirées should be held. Such proposals have been made at intervals ever since and have indeed led to certain action but not to any settled tradition; every time they are mentioned there is a sense of *déjà vu* for a good many individuals.

Library Committee

The Society's library with its collections of ancient and modern astronomical works and its archive of the papers and observations of historically significant astronomers, is an important national scientific asset. In recent years this has been publicly recognized by the willingness of official bodies and private foundations to help the Society to provide for its proper maintenance. But over the whole past lifetime of the Society the library has been built up to its status as one of the greatest astronomical libraries in the world by gifts and bequests from Fellows, by the generosity of observatories and other institutions in presenting publications or the exchange of the Society's own publications with such bodies, by the conscientious management of successive Library Committees, and by the loyal care of the Society's staff. The committee was not uniformly expert in its business but it was usually prudent and invariably devoted. Up to long after World War II, the staff included no-one with any training in librarianship, and not even anyone working full time in the library. So the fact that in the times here considered it was in such reasonably good order is both a mystery and a miracle.

The committee consisted of seven Fellows; it elected its own Chairman, those in the decade being J. Jackson (Secretary RAS 1924–29, President RAS 1953–55) 1931–33, who was elected following the death in 1930 of the long-serving H.H. Turner; W.M. Smart 1933–37; J.H. Reynolds 1937–38, 1939–40; F.W. Dyson 1938. Throughout the period the Clerk, Miss Edna Wadsworth performed the duties of librarian in addition to many others. The Council paid for occasional temporary assistance in particular tasks.

The committee met at least twice a year, and it set up an occasional subcommittee to advise on non-routine business. The routine business concerned the purchase of books and binding of journals, which it could carry out within its minute budget of about £130 a year (apart from salary). Any other expediture could be made only by reference to the Council, e.g. in 1938 about £200 for shelving in the Spencer Room — provided out of the Spencer bequest, hence the name. It advised the Council upon exchange of publications, there being many more exchanges in those days than before or since, one believes.

It naturally dealt also with such matters as:
(a)   Recataloguing the library. Shortly before the war this was declared

to be urgently needed, but then it had to wait until some years after the war. It was not only cataloguing that was needed but also a checking of the shelves; in 1933 it was discovered that the Society possessed 'a copy of the excessively rare edition of Kepler's *Tabulae Rudolphinae* containing Kepler's World Map' — it had been catalogued erroneously. In 1934 there was discovered at the back of a shelf in the upper library the Society's long-lost copy of the great folio volume of the complete works of Archimedes in the Greek and Latin edition of Joseph Torelli, published in 1792 in Oxford by the Clarendon Press.

(b) Proper storage of archival material. Considering the rudimentary and amateurish methods actually adopted at the time, it is a wonder that there seems never to have been serious loss or damage, but it is no wonder that more recent librarians and an archivist have found items of great interest that had been forgotten since those times.

(c) Precautions for the wartime care of the library, particularly the most valuable items — see next chapter.

Two particular undertakings absorbed much of the committee's energy from the end of 1934 onwards. The first arose as a result of the munificent bequest in 1922 of the Grove-Hills Library of ancient mathematical and astronomical works. This was noted in the Preface to the first volume of the *History* and in the preceding chapter of the present volume. Very naturally the Society already possessed copies of some notable items included in this collection — to the number of about 50. At its meeting in 1934 November, the committee resolved in principle, subject to Council approval, to offer for sale one copy of each duplicated work. The aim was to accumulate a fund which it could use, as opportunity arose, to acquire other rare books of which the Society possessed no copy. The principle may have apeared to be sound commonsense but its application turned out to be disastrous. This was partly because there was no ready market in those years and partly because the Society then had no professional librarian and no member of the committee seems to have had experience in such dealings. Instead of letting it be known that a relatively large number of items were on offer, it would have been far better to wait until the committee had discovered the availability of some book it particularly desired for the library. It could have borrowed the purchase price from the Society' general fund and in due course realized on whatever items of its own it was most profitable to do at the time. In letting things go for what they could get for them, for instance they sold Hevelius *Selenographia* 1647 for five guineas, and they sold the first Greek edition of Euclid's works to the Mathematical Association for five pounds. At the time of writing (1982) these items would be expected to fetch sums of the order of a hundred times greater. Years later, after the death of the purchaser of the Hevelius book, someone noticed the Society's name therein and returned it on the natural assumption that it had simply been borrowed from the library. When the librarian of that later time discovered and reported the facts, the Society bought back the book.

Another case was the first edition of Copernicus *De revolutionibus* with the Spitalfields Mathematical Society stamp, which had come to the RAS in 1846. In 1936 May the RAS sold it as a duplicate to one Dr R.S. Clay for £120. In 1954, after it had passed through various hands, the University of Illinois bought it from a London bookseller for £850. It has been missing from the University of Illinois Library since 1977. (The writer is indebted to Professor O. Gingerich of Harvard University for these details.)

The second special undertaking was a weeding out of the main library. In 1936 a subcommittee inspected the shelves and, after reference to the Council, it drew up: List A of books that should be no longer retained and should be offered gratis to Fellows or others; List B of certain duplicate books to be disposed of similarly; List C of duplicates of books in the main library, as well as in the Grove-Hills Library, that should be offered for sale to the highest bidder, first amongst Fellows and then non-Fellows. Most of the books offered gratis were taken up but some were eventually destroyed. Over the next two or three years there was much difficulty in selling most of the duplicate books; it is not clear from the records whether the attempt had been abandoned before the outbreak of war. Up to the end of 1939 the 'duplicate book fund' had indeed been used for one or two purchases but it stood in the Treasurer's accounts at a mere £67. Also in the same operation a few items connected with the Spitalfields Mathematical Society, which were not of astronomical interest, were given to the London Mathematical Society.

As against this tale of frustration, in *Monthly Notices* **100**, 334–341 there are listed for the year 1939 nearly 90 'public institutions and persons who have contributed to the Library, etc., during the year'; over 240 'Institutions, etc.' with which the Society exchanged publications; just 11 periodicals that were purchased. So it is plain where the action was. But of course it was the receipt of so much material by gift and exchange that drove the Committee to seek ways of clearing some of the shelves.

Photographic Committee

The Photographic Committee also consisted of seven members. In the decade its Chairmen were J.H. Reynolds 1931–35 and C.R. Davidson FRS (1875–1970) 1935–40; the Secretary was P.J. Melotte (1880–1961) (Jackson Gwilt Medallist 1909); who served from 1913 to 1950 with untiring dedication and devotion. It was very hard working, on a scale never to be resumed in later years; it met 50 times in the decade.

From before the turn of the century (see the first volume of the

History pp.214–15) a notable service of the Society to the astronomical community was to make available at low cost the best possible reproductions, in the form of slides or prints, of original astronomical photographs from observatories all over the world. Apart from time-determination and parts of positional astronomy, in the years when astronomy meant optical astronomy, most of the observatories existed for the purpose of securing these photographs of celestial objects or their spectra. In general, reproductions naturally could not serve as 'research material' but they did make the astronomy of the times unique among sciences in letting it be widely seen just what such material looked like.

At any one time the Society had available several hundred different subjects. The committee's principal routine task was to procure new subjects so as to keep the collection up to date, to obtain better photographs of old subjects, to employ a commercial firm to make the reproductions, to inspect all its work as received, to provide an up-to-date catalogue and to supervise all the business dealing involved. In the eight pre-war years 1931–38 the Society sold about 5000 slides and about 3000 prints at a price of one shilling (5p) each — the price rose a little in 1939. As a service to astronomy this shows it to have been on a worthwhile scale. It was never intended to bring in a financial gain; the annual turnover in the routine working of the committee averaged no more than about £150, which gives no idea of the amount of work put in by its members.

Most institutions with anything to offer were only too pleased to have their work represented in the Society's collection of photographs. At this distance in time, it is amusing to note one exception. In 1932 the Chief of the United States Air Corps in Washington declined to allow the Society to reproduced a US Army aerial photograph of Meteor Crater, Arizona. It was not permissible to ask what security risk might be involved but the photograph in question had been published world-wide in the (US) *National Geographic Magazine*.

The Council entrusted various special tasks to the committee. In those days, and so long as the old Meeting Room remained in commission, photographic portraits of past Presidents used to hang in two rows along each side wall. By 1933 some older ones were becoming faded or discoloured. The Photographic Committee was required to find and employ a firm to make 'carbon' copies of these; after initial troubles, the results proved highly successful. Some more recent portraits are now (in the 1980s) to be seen on the walls of the stairway in Burlington House, but it has long been a matter of regret that the Society's meetings have no longer been conducted under the watchful gaze of every one of its former Presidents.

Between 1934 and 1937 the Committee handled the production of a

third edition of the Franklin-Adams *Chart of the Heavens*. The requirements of would-be purchasers had to be ascertained quite precisely
before the undertaking could be finally set in train, but it eventually
went through smoothly and expeditiously, with complete success for
the careful budgeting of the operation. There were about 60 purchasers
of complete or partial sets. A complete set was sold at £27, and the
whole proceeding cost about £1300; there was an estimated balance of
about £80 in favour of the Society — a rough idea of what this would
mean at 1982 rates might be had by multiplying each sum by 20. The
main work of reproduction was done by Messrs Emery Walker Ltd., and
the committee thanked them warmly for the excellence of their work.

To recall for a moment what in retrospect may seem to have been an
age of primitive visual aids, it has to be remembered that all slides were
on glass, and the projector had to be a large affair to take the standard
size of $3\frac{1}{4} \times 3\frac{1}{4}$ inches and also the American size of about $3\frac{1}{4} \times 4$
inches. The result could indeed be superior to what can now be
achieved with 35 mm plastic transparencies. There was no 'overhead
projector', greater use being made of the blackboard than in more
modern times, but there was the 'episcope' which — with luck — would
project a formula or graph drawn on an opaque surface. In 1936 the
committee acted for the Council in purchasing a Ross epidiascope for
about £50. This projected either way; so far as one is aware this was the
projector that remained in service until the use of the old Meeting Room
came to an end in 1969.

As a postscript it may be recorded that the committee met in 1939
November and recommended that slides and prints should, for the time
being, be available for purchase on meeting days only, provided voluntary help for the Assistant Secretary was forthcoming. The Photographic Committee as such was destined to hold just two more wartime
meetings, in 1940 and 1942; in 1943 it became the Photographic and
Instruments Committee, the early history of which belongs to the next
chapter of this book.

Finance Committee: the Society's affairs

Each year the Council appointed a Finance Committee composed of the
Treasurer and four or five knowledgeable senior Fellows — indeed
much the same group throughout. So long as the Society operated
normally under the conditions of the times — which of course it did
until the outbreak of war — there was very little for its Finance
Committee to do. Whatever economic problems there were for the
country or the rest of the world, for a Society like this it was a time of
financial stability, certainly compared with any subsequent decade. The
value of money changed scarcely at all, investment for such a body was

restricted to gilt-edged securities and rates of interest on these as a whole changed very little. Consequently there was little call for elaborate budgeting; the financial operation of the Society had been attained by over a century of gradual evolution, rather than by anybody's astute annual estimates. So all that was needed to keep it going was in general ordinary prudent housekeeping; if the Treasurer did much the same as last year — which was what the Society expected him to do — he knew it would cost much the same. He had, of course, to be on the lookout for the occasional perturbation and make some appropriate adjustment, and a few times during the decade in his monthly statement to the Council, the Treasurer did report a comparison of the position with that at the corresponding time in the previous year, and he based some recommendation upon it.

Over the nine year 1931-39 the annual cost of running the Society was about £3200 ± £200 and there was an average excess of income over expediture of some £24. In round figures, the cost of running the office was about £1000, the cost of printing about £1500 and of a host of items including heat and light, meeting expenses, library expenditure, etc., was about £500. The total income from Fellows' subscriptions was about £1500, that from dividends about £1000 and from sales of publications, etc., about £500. In the decade the total financial assets rose modestly from about £21000 to something over £27000. It has to be noted that the Office of Works provided the Society's premises free of rent and rates and it paid for the maintenance of the fabric and exterior decoration.

Even at the time there was some air of mystery about the meetings of the Finance Committee; allusions to them in the Council minutes were somewhat casual. There seems to have been only one direct report; in other cases the Treasurer indicated the committee's views, or there was reference to its own minutes. Unfortunately those minutes have not been traced at the time of writing this account (1982). Clearly there were no regular meetings, and the Treasurer called one only if he or the Council wished to have the committee's advice. A meeting in 1931 March and a marginal note in the Council minutes for its April meeting indicates that it pertained to the winding up of the Society's association with Messrs Wheldon and Wesley, which formerly conducted the sale of its publications, and with the disposal of money on deposit; it seems also to have advised about cases of Fellows in Australia who were then in difficulty about paying their dues because of a disadvantageous exchange rate, the Council having referred this matter to it. In 1932 June the Council empowered the committee to take action over certain investments and the indications are that it acted with promptitude. At the November Council the same year the Treasurer presented the committee's direct report mentioned above in which it called attention to the estimate of a deficit of about £400 at the end of the year (actually it was £419 when the accounts were made up several weeks later). This was because of some increase in printing costs, the cost of rewiring the premises required by the Office of Works to permit the change from direct to alternating current, and to 'a serious decrease in subscriptions from Fellows'. The committee made definite recommendations with a view to reducing printing costs. In regard to some of these the Council left it to the Secretaries to effect possible economics but it agreed to ask the Royal Society for a grant to assist with the *Geophysical Supplement* (in due course a grant of £50 was forthcoming), and it decided to discontinue the issue of free reprints to authors. The measures taken were so well-calculated that the accounts for the following year showed a surplus of £335. The Society being thus restored to an even keel, it made no further application for assistance to the Royal Society — at any rate not for many years to come — and it gradually relaxed certain economies. If in the 1980s anyone is amused by economies on this scale, he should multiply by about 20 to translate them to his own money value; he would then have the sort of figures about which *he* might worry. Also in regard to the 1980s, this account is being written in Eddington's centenary year; at the Council meeting concerned in 1931, it may be interesting to remark, Eddington proposed that an author should be limited to 40 pages in any volume of *Monthly Notices*, and should more than this be accepted for publication the author should pay the cost. Nothing came of it at the time but perhaps amongst all his other claims to fame, Eddington should be commemorated as an inventor of page charges.

The Treasurer reported to the Council a meeting of the Finance Committee held on 1937 February 9 at which it had apparently reviewed the investments portfolio and requested the Treasurer to consult a broker about certain items. After he had reported the result to the Council, it approved just a couple of changes in the portfolio.

The report of the auditors for the year 1936 mentioned some matters to do with the Society's insurance policies, including the fact that there appeared to be no inventory of the Society's property. It was this that triggered the review of the insurance of this property which is described here in the section on the House Committee. There it is mentioned that the professional valuers accepted the Council's own figure for books and stocks of publications. That figure resulted from a joint meeting of the Finance Committee and the Library Committee held on 1937 February 23.

The only other meetings of the Finance Commitee in the decade seem to be one in 1939 June and one in 1940 March or April, both to consider the investment of particular funds as requested by the Council, which on the second occasion had given the committee power to act.

During the decade the Society gratefully received several bequests of which it was able to make effective use. Viewed in comparison with the sums involved in the day-to-day running of the Society the amounts of these legacies were highly significant, although a later generation might have difficulty in appreciating this and might be surprised by the trouble taken by the Council in handling these moneys. From time to time the Society still publishes some particulars of such past benefactions, so that it is unnecessary to record many details here.

In 1931 A.F. Lindemann, the father of F.A. Lindemann (Lord Cherwell), bequeathed £1000 which the Society was free to use for its general purposes; in 1937 Miss Julia Lindley made a similar bequest also of £1000. In 1933 the Society received about £300 as a share of the residuary estate of archdeacon Beresford Potter, which it was required to invest; the Council allocated the income for the purposes of the library. In 1936 Captain J.J.L. Goodridge bequeathed £50, which likewise had to be invested. In 1936 A.G. Stillhamer, a physicist and astronomer of Bloomington, Illinois, a Fellow of the Society, died leaving his estate to Miss M.S. Capps of that city; she declared she was fulfilling a wish he had expressed by presenting to the Society the sum of $2500 to form the A.G. Stillhamer Trust, the income to be used for scientific research or instruction.

In 1936 the Society was one of twelve to benefit in the eventual winding up of the estate of the social philosopher Herbert Spencer (1820–1903); its share amounted to £1241. The condition of acceptance was that the sum should be spent in specified ways over the next five years, no part being used for endowment. The Council used some part for bookshelves and other furnishing in what then became the 'Spencer Room' in the library. It also invited applications from astronomers for grants in accordance with the condition and in 1939 April it set up a strong committee of seven members to consider them. In due course this committee recommended a number of awards; some of these had to be made conditional upon the recipients obtaining other support as well, and when this did not materialize the award could not be confirmed. As examples that did have happy outcomes, one was £100 awarded to E.H. Collinson and J.P.M. Prentice for Schmidt cameras and other equipment for the photography of meteors — the Council going out of its way to welcome the opportunity to support such work — and another was £50 to W.H. Steavenson towards the erection of an 'observatory' to house his 30-inch reflecting telescope on a site provided in the grounds of the Cambridge Observatories. One application that was not approved at the time was from the editors of *The Observatory* for £100 to help that magazine to survive the coming emergency. The committee had been sympathetic, but it thought that the application should be referred to a more representative committee. In due course (1939 December 8 Minute 214) the Council appointed such a committee consisting of the original seven with another twelve and with the editors invited to attend. This was a ludicrous number and there is no record that it was ever called into action. However, the seed had been sown and the Society a year or two later embarked upon a scheme of support for *The Observatory* that has proved effective ever since.

E.W. Brown (1866−1938) (Gold Medal 1907), an Englishman long associated with Yale University in the United States, left to the Society the sum of $5000, the income to be used to assist in the calculation of accurate orbits of solar system bodies and any not so used to be applied to the general purposes of the Society.

The eminent amateur astronomer A. Stanley Williams (1861−1938) (Jackson Gwilt Medal 1923) bequeathed to the Society all his astronomical instruments, books and papers, together with the sum of £200 for reduction or publication of his observations of variable stars, and also a legacy of £100 for the Council to use as it saw fit. In 1939 April the Council appointed a committee to advise upon the treatment of the observations and in due course, on the committee's recommendation, it asked P.M. Ryves to make a detailed examination and to report. Unfortunately the war delayed further progress for a long time after that.

Among gifts not concerned particularly with finance or the library may be mentioned: In 1931 J.J. Hall (1845−1941) a Fellow since 1899 and well-known horologist presented a gold pocket chronometer (Padbury 3474) which thenceforth until the war was placed on the President's table at meetings of the Society. It was he who had also presented to the Society the Arnold chronometer used in A.J. Carroll's 1936 eclipse expedition to Siberia. Also in 1931 Miss F. Herschel presented a portrait of Francisca Herschel (1846−1932) taken by Sir W.J. Herschel, Bart, in 1916 by the process of Lumière's colour photography. In 1936 Dr W.A. Parr, past President and Honorary Librarian of the British Astronomical Association, presented a small bronze bust of Galileo, still to be seen in the library. In 1936 Dr Otto Boeddicker presented the original drawing of the Milky Way which he had made over the years 1884−89 during the time that he was assistant to the fourth Earl of Rosse in his observatory at Birr Castle in Ireland (see *Monthly Notices* **96**, 641−642, 1936).

One 'gift' deserves special mention. At one meeting the Council caused its thanks to be conveyed to one very eminent astronomer for a 'remarkably fine prism' which had been received from him. At the next meeting it heard from him that he had simply been returning some of the Society's property that he had had on loan.

House Committee

The House Committee comprised five to seven members. The Treasurer was customarily one of these and he normally took the Chair at its meetings. In 1931 May the Committee elected W.H. Steavenson (1894–1975) (President RAS 1957–59) as its secretary and he served as such for the rest of the decade. The committee advised the Council upon everything to do with the maintenance of the Society's premises, upon all domestic arrangements and upon the care of all its belongings apart from those expressly within the purview of the Library or Photographic Committees. Naturally a certain amount of consultation between committees had to take place but this could generally be operated through the Treasurer and Secretaries, without the need for formal exchanges.

The Society occupied the quarters in Burlington House into which it had moved in 1874. Since at the time of writing the old Meeting Room as such is no longer to be seen, it may be of interest to recall that it occupied the whole of the ground floor on the north side of the entrance hall. The platform was at the western end; the seating was tiered and the layout worked well for seeing and hearing the proceedings. Until it was removed in 1936 in order to make space for the epidiascope and an operator, there was in the centre of the north wall a fireplace with a large marble mantlepiece. It was on the latter that the certificates of candidates for Fellowship used to be 'suspended'. The ventilation of this room, and that of the Council Room, was a perennial problem; it was never adequately solved and it may be presumed that in the old days fires in open fireplaces operated better than any sophisticated devices of more modern times.

Together with the Royal Society and several other learned societies, the Society held its accommodation from the Government free of rent and rates. HM Office of Works was in charge of Burlington House; it was responsible for the fabric and the exterior decoration and its approval was required for any structural modifications. The Society was responsible for interior decoration and, of course, for heating and lighting charges.

Over the years up to 1931 the Society had acquired by gift and bequest a few hundred astronomical instruments of every conceivable sort. Unfortunately the first volume of the *History* included no systematic account of those received up to 1920 nor, in particular, of the fates of any not retained in the possession of the Society. Happily the history of the whole subject has been written up by H.D. Howse, *Q.Jl R. astr. Soc.* **27**, 212–226, 1986, who has recorded the fate of practically every instrument ever to come into the Society's hands. The stewardship of all this equipment was a highly significant element in the Society's service to both professional and amateur astronomers.

The Report of the Council to the Annual General Meeting of 1931 February 13 contains the 'Report on the Society's instruments' prepared at the Council's request by W.H. Steavenson (*Mon. Not. R. astr. Soc.* **91**, 310–311, 1931). It applied to about 150 items; he had classified the instruments under those that he proposed the Society should: I keep for use on the premises or on loan; II, III lend/give to museums; IV, V give to educational institutions/small astronomical societies; VI dispose of as being no longer of scientific value. The meeting empowered the Council to proceed in this way, and the auditors for 1931 reported to the next Annual General Meeting that a number of the instruments had been dealt with accordingly. Almost 20 went to the Science Museum, South Kensington, including (on loan) a Herschel 7-foot reflector, a Cavendish Apparatus, the South Transit Instrument and a sectant by Bird that had belonged to Captain James Cook RN, as well as certain items from the 'Sheepshanks collection'; the Science Museum has received a few others of the Society's instruments before and since this operation of 1931. About 60 items, again including some 'Sheepshanks' instruments, went to the Museum of the History of Science, Oxford, then known as the Lewis Evans Collection. No further disposal of instruments took place before the war; each year every borrower of any instrument was required to apply for a renewal of the loan, or to return the instrument to the Society. All this action arose out of an inspection of the instruments that had been made by the House Committee in 1930 November, and was the chief concern of the committee throughout 1931.

In the few years after 1931 the committee met at somewhat irregular intervals, probably on account of the miscellaneous character of the business which it had to handle: rewiring of the premises, purchasing easy chairs for the library and an adding-machine for the office, attending to minor repairs and maintenance, installation of a house telephone, and so forth.

Then in 1937 the Council sought the committee's opinion upon a professional valuation of the Society's property which it had obtained for insurance purposes. The committee suggested nominal figures of £100 for instruments and £100 for Presidents' portraits. This was presumably because, if any of these were lost, there would be nothing the Society could buy to replace it; in those days the Society seems to have regarded its property as being for use and enjoyment, not at all as investments. Apparently the professionals accepted as reasonable the Council's own figure of about £12000 for all its books as well as stocks of publications. So, with the professionals' own figure of about £3000 for furnishings, pictures, clocks, etc, the committee recommended the Council to accept these valuations and to insure for a total of about £15000. (The interest in quoting this is mainly to give a basis for a measure of subsequent inflation.)

While mentioning money matters in which the committee was concerned, one must remark upon the extreme care and prudence which it exercised. For one job of work that eventually cost less than £200, it obtained estimates from no fewer than eight different firms. Also when the state of health of the porter's wife made it necessary to engage some extra help in cleaning the premises at the cost of five shillings a week, it came to an agreement with her and her husband to deduct from her weekly wage the sum of two shillings and sixpence (12½p). Incidentally, the individual in question survived for almost another 50 years, until 1980; by paying its share of her pension right up to the end, the Society may be said to have atoned for that small economy.

Early in 1938 the committee started planning the work of redecorating and recarpeting the hall and staircase, and of reconditioning and furnishing the Spencer Room. It finally approved a colour scheme on June 30; three months later it inspected the finished work and reported that it had been satisfactorily carried out. This was less than a year before the eventual outbreak of war but as early as 1939 January the Council asked the committee to consider what should be done to protect the Society's property in the event of war. It proceeded to consult HM Office of Works about reinforcing a room in the basement. However, in the following March, the Office of Works having declared its inability to undertake immediate work of this sort, the committee decided that the Society ought to proceed independently of it, and of the other societies in Burlington House. It recommended that the Treasurer should arrange to have the most valuable property stored outside London, preferably in Oxford, and less valuable material stored in a strengthened room in the basement, that should be made ready on the Society's own initiative. The Council did in fact proceed accordingly but as war loomed ever nearer the necessary action had to be taken mainly by the Officers on their own. The House Committee did not meet again until 1945 December 13.

Geophysical Committee

The geophysical activities of the Society prospered throughout the decade, following what had become a well-established pattern. They were organized by the, in practice almost autonomous, Geophysical Committee. This was quite a large body, of between 20 and 30 members, including most of the established geophysicists in the country, together with some eminent meteorologists and geologists, as well as physicists having an interest in possible geophysical applications of their work and one or two interested professional astronomers. The Council appointed nine members, and the rest were appointed by outside bodies. It was considered that by accepting membership such scientists made themselves available for consultation about the committee's work, even though some might rarely be able to attend its meetings. The Council appointed the Secretary (see 'Geophysical Secretary' earlier in the chapter), who was W.M.H. Greaves 1927–31 and R. Stoneley 1931–45. The committee elected its own Chairman: Sir Gilbert Walker FRS (1868–1958), 1931–36; S. Chapman FRS (1888–1970), (President RAS 1941–43; Gold Medal 1949) 1936–39; (Sir) David Brunt FRS (1886–1965), 1939–51. It met once a year, in the spring, and its business then was almost exclusively to arrange the programme of geophysical meetings for the following session — occasionally it considered also matters to do with the *Geophysical Supplement.*

The committee arranged six 'Geophysical Discussions' for the session, normally to be held on the last Friday of the months between October and May. Two of these were for the 'reading' of papers submitted for publication in the *Supplement.* Each of the other four was a discussion of a particular topic agreed upon by the committee, which in general proposed the speaker who should be invited to open the discussion and maybe suggested who should be invited to preside. However, the detailed arrangements were all left to the Secretary. All kinds of distinguished scientists accepted invitations to take part in these ways. The proceedings were conducted on an agreeably informal basis, with, in general, no prearranged programme of speakers after the opener of the discussion but there was never a shortage of participants. Typical topics and opening speakers during the years concerned were: 'Light of the night sky with special reference to its spectrum', Lord Rayleigh; 'Isostasy', Sir Gerald Lenox-Conyngham; 'Upper air ionisation', (Sir) Edward Appleton; 'Solar–terrestrial relationships, S. Chapman; 'Lightning', (Sir) Basil Schonland; 'Ice Ages', Sir George Simpson; 'Topography of the Ocean Floor', Lt Col. R.B.S. Sewell.

Occasionally meetings were organized jointly with the Geological Society, for example: 1933 January 25 in the rooms of the Geological Society on 'The development of fractures in the Earth's crust'; 1934 January 26 in the rooms of the Royal Astronomical Society on 'The origin of the Earth's major surface features'. This latter meeting was followed by a dinner at Stewart's Restaurant, Old Bond Street, presided over by F.J.M. Stratton, President RAS. The meeting itself was a notable success, with an audience of about 140, but only 25 attended the dinner (even though the cost per head was only four shillings and sixpence, excluding wine).

At the reading of papers the average attendance was fewer than 20; at the ordinary discussions it ranged from 30 to 70; at the joint meetings it was between 100 and 150. In 1937 the Secretary (Stoneley) remarked that the average attendance had changed little since 1920. These numbers were fairly modest but the foundations were being laid for the great expansion of geophysical science after World War II. It should also be noted that the individuals concerned in these activities included several of the very greatest pioneers in the field. The Society has cause for gratification in what it was able to do to foster this science in its formative years, but the part played about the same time, particularly in regard to seismology, by the British Association should also be remembered.

One fairly early sign of the growth that was to come was that at the meeting of the committee in 1933 March H. Jeffreys proposed that the *Geophysical Supplement* should appear three times (instead of twice) a year; it was agreed to leave this to the Chairman and Secretary. Actually the rate of three issues a year was not attained until 1939.

During the years concerned, the Society's geophysical activities had more links with the armed services, and with Government establishments like the Meteorological Office, certainly than has been the case since the various Research Councils were set up. Serving officers of the Royal Engineers and of the Indian Army frequently contributed to the Discussions. A highlight of co-operation of this general character originated from the meeting of the committtee in 1937 March. It agreed to a resolution proposed by O.T. Jones, seconded by H. Jeffreys, 'That it is desirable that steps should be taken to test by seismic methods the depth of unconsolidated sediments that underlie the continental shelf off the coasts of the British Isles.' The committee agreed to inform the Astronomer Royal and the Hydrographer of the Royal Navy about this. The full significance of the question is not immediately evident today but it must have been a crucial issue at the time, for the resolution provoked an energetic response. At the next meeting of the committee in 1937 May, S. Chapman reported that the Society had transmitted the resolution to the Royal Society. After reference to its National Committee for Geodesy and Geophysics, that Society had set up a committee with the Hydrographer as Chairman and had voted £500 for the work. The Admiralty had provided a survey ship and motor boats for seismic exploration of the ocean floor, together with a submarine to make gravity measurements using apparatus lent by the Danish geophysicist Vening Meinesz — an addition suggested by the National Committee. The work was planned by Sir Gerald Lenox-Conyngham, (Sir) Edward Bullard and B.C. Browne. At a Discussion held on 1939 March 24 presided over by the Hydrographer Rear Admiral J.A. Edgell RN FRS, these geophysicists were able to present the results of the successful execution of the whole operation.

In 1939 the normal programme of meetings went on until May, when the committee had its usual annual meeting and, as usual, discussed the programme for the coming session. That meeting was presided over by

(Sir) David Brunt; its minutes were later approved and signed by 'D. Brunt' on 1945 October 26. Not much of the programme that had been discussed in 1939 had ever materialized but there was one discussion on 1940 May 24, with an attendance of 23, on 'River Flow' introduced by Captain W.N. McClean. After that there were only three other Discussions and no meeting of the committee until the war was over; but these were in the next decade.

Board of Visitors of the Royal Observatory, Greenwich

In 1710 Queen Anne appointed certain Fellows of the Royal Society to serve as Visitors of the Royal Observatory. By Royal Warrant each successive Sovereign re-appointed such Visitors. In 1830 the Warrant of William IV set up a reconstituted Board of Visitors composed of the Presidents and five other Fellows of the Royal Society and of the Royal Astronomical Society (together with the professors of Astronomy in Oxford and in Cambridge and (later) the Hydrographer of the Royal Navy). The term of service of each ordinary Fellow serving as a Visitor was five years. It was thus the duty of the President of the Society each year to appoint, generally, one new Visitor different from the retiring Visitor, although he might, and sometimes did, appoint one who had served an earlier term. In the years concerned here, he acted entirely on his own discretion although it was considered proper that he should inform the Council of his action. In these years, this Board and the JPEC (see below) were the only outside bodies on which the Society was officially represented. The first volume of the *History* contains only one brief mention of the Visitors but there is a good account of their whole story by P.S. Laurie 'The Board of Visitors of the Royal Observatory', *Q. Jl R. astr. Soc.* **7**, 169–185 1966 and **8**, 335–353 1967. This was written after Queen Elizabeth II had dissolved the Board by a Royal Warrant signed on 1965 August 27 when the control of the Royal Observatory had passed to the newly established Science Research Council.

Joint Permanent Eclipse Committee

The origin of the Joint Permanent Eclipse Committee (JPEC — 'joint' between the Society and the Royal Society) was recorded in the first volume of the *History* (pp.215–216). At the time here concerned, the Society each year appointed or re-appointed eleven representatives and the Royal Society appointed eleven of its own Fellows. It elected its own Chairman and Secretary; F.J.M. Stratton had become Secretary in the preceding decade and he continued to serve for 32 years.

The commitee was very active in its intended function of organizing expeditions to observe total eclipses of the Sun from various parts of the globe and in all necessary preparations, even to the design and construction of special instruments. Its own key role was, of course, to investigate the financial requirements, to apply to the Government Grant Committee of the Royal Society for the money and to administer its disbursement. Through the years it had accumulated a considerable amount of specialized equipment, over which it retained general control, although it was only too glad for it to be used for any other scientific purposes when not required for eclipse work. The new expenditure on instruments for a particular eclipse was typically a few hundred pounds and that on an actual expedition (transport, subsistence, etc.) was of the order of a thousand pounds. The examination of proposed investigations was so thorough and the budgeting so meticulous that, so far as one can discover, the Grants Committee always gave the amount applied for. Only once did it propose any condition of its own; it asked for certain observations on the ionosphere to be made through the time of the eclipse of 1936 June. On that occasion the JPEC found that such observations were already planned by an expedition under other auspices and it decided not to try to duplicate them, however it did seek to organize ionospheric work on a subsequent occasion.

The committee's judgement was trusted with such respect because it was indeed very high-powered. It comprised the Astronomer Royal, the most experienced British eclipse observers of the time, representatives of institutions likely to participate in expeditions, and one or two highly regarded theorists with an interest in solar physics. It would involve too much detail to list all the names but as just one example, the JPEC minutes of the meeting on 1932 December 9 take special note of the retirement of Sir Arthur Schuster FRS (1851-1934), a great pioneer of modern astrophysics, who had been a member since the committee was first established.

Eclipse-chasing seems always to have had a particular appeal for British astronomers. Many of them cheerfully endured extreme rigours of travel and climate in order to erect their telescopes on sites that were normally regarded as quite inaccessible. The grim seriousness of the JPEC deliberations was usually not permitted to betray the spirit of adventure that inspired them but one touch of humour did manage to invade the minutes of 1934 April 13 when, it is recorded, the committee decided to enquire of the Admiralty about Sarah Anne Island 'formerly in the belt of totality of the 1937 eclipse and now reported missing.' The next meeting on 1934 December 15 heard simply that the Hydrographer of the Royal Navy had written about the 'non-existence' of the island — evidently someone felt slightly embarrassed at having been not quite solemn enough about a no-joking matter, to wit, the demise of a Pacific island.

Anyone in Britain wishing to mount a serious eclipse expedition was only too glad to work with and through the JPEC. It provided access to the best scientific advice and instrumentation, and not only direct financial assistance but its prestige won diplomatic advantages and material aid; for instance, in 1936 the Blue Funnel Line carried 12 tons of equipment to Japan and back without charge.

In the decade, the JPEC concentrated its efforts upon the eclipses of 1932 August 31, 1936 June 19, 1940 October 1. In each case the preparations occupied most of the committee's attention, as well as that of an occasional subcommittee, for about the preceding four years.

In 1932 expeditions went to three sites in Canada. Unfortunately two were completely clouded out on the day. One from the Royal Observatory was somewhat handicapped by cloud, but it did achieve some valuable observations, which were duly published in *Monthly Notices*.

The 1936 eclipse was perhaps the high point of eclipse observing of the Sun itself — as distinct from the testing of Einstein's 'bending' of light first carried out at the famous eclipse of 1919. Well-equipped British expeditions went to Omsk in Southern Siberia, where they met with great success, and to Hokkaido in Japan, where they obtained valuable results, although hampered by cloud. The Society accepted the invitation of the Royal Society to hold a joint meeting on 1937 May 27 when all the results were presented and discussed.

Plans for observing the 1940 eclipse from Brazil and South Africa were well advanced before the war but early in 1940 the JPEC had finally to decide that no direct British participation would be possible. It did give what support it could to the available astronomers in South Africa at the time.

The history of eclipse observing in the first half of the century is a tale of much frustration by war and weather. Some observers achieved essential results but for others there was nothing to show for enormous efforts in preparations. Also there were bound to be diminishing returns in the way of new discoveries. So by way of postscript it may be recorded here that the two societies agreed to the dissolution of the JPEC in 1970. British astronomers will continue to observe eclipses but they now seek support through more routine channels. The romance that so long attached to the doings of the JPEC obviously remains in astronomy, but much of it has passed to other branches. There is 'an outline of J.P.E.C. activities' in the *Quarterly Journal* **12**, 39–44, 1971 by R.O. Redman who had been a member from 1949, was long its Secretary, then Chairman in its final year.

Publications

Throughout the decade the Society 'published and sold' its own regular journals. For many years up to 1929 these had been sold for the Society by Messrs Wheldon and Wesley; the business of winding up that arrangement dragged on from then until well into 1931 (Council Minute 18 of 1931 November 13). Until 1931 the general style of printing *Monthly Notices* had changed rather little since its earliest days; the *Geophysical Supplement*, which had been started in 1922 as literally a supplement to *Monthly Notices*, was naturally produced in the same style. In particular, the page size was smaller than had become the fashion for many professional scientific journals. At the Council meeting of 1931 April 10 (minutes, p. 178) H. Dingle gave notice of a motion 'that the sizes of page and print of the *Monthly Notices* be altered to those of the *Proceedings* of the Royal Society.' After consideration at the next two meetings, at its June meeting Council resolved that from *Monthly Notices* volume 92 and *Geophysical Supplement* volume 3 the size of the page be that of the *Proceedings* of the Royal Society. It appointed what quickly became known as the Type Committee 'to consider possible forms of type and to recommend one which...will not involve a total increase of cost per volume of the *Monthly Notices* of more than about £50 as compared with the present cost.' At the time the cost of printing the annual volume was running at about £1000. The committee consisted of the Officers and L.J. Comrie, Superintendent of HM Nautical Almanac Office 1931-36, who in the course of his official duties and in his activity in producing mathematical tables had made himself an expert in the sort of printing involved. It held two main meetings in 1931 July; it made detailed technical recommendations of its own, which it was well-qualified to do, and discussed samples of work produced in accordance with these by the Society's printers, Messrs Neill & Co. Ltd., of Edinburgh. It revised some of its recommendations in consideration of comments from members of Council and others. As a result, *Monthly Notices* appeared in its new format in 1931 November, and a very high class format it was, until, after about ten years, wartime exigencies called for various restraints and economies.

In the meantime, in 1937, when a problem arose about the lettering of diagrams in the Society's publications, Council set up a Diagrams Committee. It held three meetings in 1937 September-November; besides drawing up recommendations about the preparation of diagrams and making of blocks, for some reason which is now obscure, it also played with the idea of changing to another printer. Eventually, however, it advised continuing with Neill & Co., but with some adjustments in the terms of their contract. The impression one receives from looking back at the records is that the Society drove a hard bargain. Nevertheless relations with Neill & Co. continued to be friendly; in 1940 January 12 the Council decided to send a letter of thanks to their retiring Managing Director for the interest he had shown in the Society's publications over a long period of years.

In the first volume of the *History* (pp.228 and 239–243) J.L.E. Dreyer gave an interesting sketch of the history of the Society's publications up to 1920, and the excellent index to that volume contains references to numerous details. The main publications followed much the same pattern from 1920 through the decade with which we are concerned. Since, however, the pattern has changed so considerably in more recent decades, it may be of interest here to recall the character of those publications during the 1930s.

The *Monthly Notices* were associated with the months in the Society's session. Consequently, number 1 of a volume reported the proceedings of the November Ordinary Meeting and for the rest it was intended to consist of all the papers formally accepted by the Council on the day of that meeting, most of these having been 'read' at the meeting. It was scheduled to be published within six or seven weeks of the meeting. The printing contract gave dates for each stage: proofs, page proofs, printing and distribution. At their session on the Saturday morning after the meeting, while one of the Secretaries attended to the Council minutes and the business arising therefrom, the other Secretary 'edited' all the accepted papers. Since the Secretaries would have seen that the copy was in pretty good shape before proposing its acceptance by the Council, presumably this editing was mainly the insertion of necessary final instructions for the printer. The bulk of the copy was then mailed to the printer that day and he would have it to start work on by the Monday morning. The chief problem in keeping to the schedule was held to be that of getting proofs back from authors.

Numbers 1–3, 5–8 applied in the same way to the months up to June, with the exception of February. The volume was completed by number 9, labelled 'Supplementary number' because it was associated with no particular meeting, but contained papers accepted during the summer recess; it was published during October. Number 4 was the February issue; besides reporting the Annual General Meeting held that month, including the election of the new Council which took office at the end of that meeting, it contained the Report of the retiring Council for the preceding calendar year which had been formally presented, and was indeed printed by resolution of the meeting.

The Report year-by-year was a repository of much of the history of astronomy, as well as of the Society itself. As already mentioned, at the December meeting the Society appointed two of its Fellows to serve as 'auditors of the Treasurer's accounts' for the year then coming to its end. Their report was the first item in the number; it included the audited accounts, some comments thereon as well as the results of their inspection of the Society's premises and property, and the traditional table of the 'Progress and present state of the Society'. All of it was the outcome of strenuous exertions on the part of the auditors. This was

the case particularly as regards the innocent-looking table. Even though
the office kept its records with meticulous care, and even though
perhaps 90 per cent of the Fellows were conscientious about paying
their dues, it was always astonishingly difficult to be precise about the
membership at a given date because of the residue of Fellows who
forgot to pay, or how much to pay, or to report a change of address, or
who had difficulty in transmitting money from abroad, or who died
without the Society being informed. So, besides presenting their table,
an important duty of the auditors was to lay before the President a list
of Fellows in arrears with their payments, with the amount due from
each. Of course this list was not published; but in due course the list of
'six-guinea defaulters' was suspended in the Meeting Room. The Treasurer had to take what action he could and to report to the Council;
individuals who did not make a satisfactory response in accordance
with the bye-laws then ceased to be Fellows; and this was taken into
account when the following year's auditors drew up their table.

All this was followed by a list of the chief financial gifts to the
Society ever since its foundation and by somewhat needlessly detailed
accounts of all the special funds and trust funds resulting from such
gifts. No doubt there was some element of hope that other benefactors
might be thereby encouraged. Then there were short reports on celestial photographs on sale to Fellows, on the Society's publications over
the past year, and on its instruments. Later in the Council's Report there
was a list of those who had contributed to the library during the year
and of institutions with which the Society had exchanged publications.
In the Report for 1939, for instance, these lists occupied eight printed
pages. And here it ought also to be recalled that with each issue of
Monthly Notices there came a printed list of the latest additions to the
library.

There were obituary notices of almost all the Fellows and Associates whose deaths had been reported during the year. They were
printed in an unpretentious style, shorter ones being in general un-
signed, others being initialled only up to 1938 but signed from 1939.
Some devoted — and very busy — Fellows were ready to contribute
such notices whenever called upon; F.W. Dyson, H.P. Hollis and H.C.
Plummer being some examples of those who were particularly generous in this rather melancholy regard. Actually the manner of production
rendered these notices more spontaneous than more formal tributes
and they contain many interesting and agreeable sidelights upon personalities and events in the history of astronomy.

A section on 'Proceedings of observatories' consisted of reports on
the year's work supplied by the Directors of the Royal Observatories,
university observatories, Commonwealth and certain other overseas
observatories, and reports of numerous private observatories operated

by amateurs, some of them very notable personalities in the astronomical community.

There followed the famous 'Reports on the progress of astronomy', commonly known simply as 'Council notes'. Much deliberation amongst the Officers and Council, much not recorded in any minutes, went into the choice of topics and authors, and it was only from 1938 that the authors' names were spelled out, although the owners of the initials used till then were easily identifiable. Some 'Notes' were isolated reviews of particular developments possessing topical interest, for example: Discovery of Pluto, by A.C.D.C(rommelin) (**91**, 380−385, 1931); Expansion of the Universe, by A.S.E(ddington) (**91**, 412−416, 1931); Stellar Motions, by J.H. Oort (**99**, 369−384, 1939); Seismological tables, by Harold Jeffreys (**99**, 397−408, 1939); Theory of errors by H.R. Hulme (**100**, 303−314, 1940). Other subjects were dealt with regularly every year or every few years, e.g. comets, solar activity, double stars, extragalactic nebulae, and so on. In any such case, often the same author contributed the 'Note' for many years; for instance, A.C.D. Crommelin wrote on minor planets and comets for over 40 years and H.W. Newton wrote on solar activity for 36 years. As a whole, these Reports were the forerunners of the sort of volumes of annual reviews that started to appear elsewhere in the 1960s, but nothing seems quite to have taken the place of the regular items on comets, etc.

Finally, *Monthly Notices* number 4 contained the written versions of the President's Address(es) delivered at the Anniversary Meeting, as described in the section dealing with that meeting.

In 1934 the Society published *General Index to the Monthly Notices volumes 71−91 (1911−1931)*, and in 1953 it published *General Index to the Monthly Notices 92−110 (1932−1950), Geophysical Supplement 1−5 (1922−1949), Memoirs 61−66 (1917−1950), Occasional Notes Nos. 1−12 (1938−1950)*. So the first *Index* was produced in the decade, while the second *Index* covers most of the publications in the decade. There were ten volumes of *Monthly Notices* within the decade, averaging about 800 pages a volume, *Geophysical Supplement* volume 2, No. 7 to volume 4, No. 8, *Memoirs*, parts of volumes 65 and 66 and *Occasional Notes* Nos. 1−8. Also a full list of Fellows and Associates was published every second year up to 1939. A catalogue of slides and prints was published from time to time.

This brings us to the story of *Occasional Notes*, which is another example of the spirit of prompt enterprise within available means, of those who conducted the Society's affairs in the times under review. At its meeting of 1938 February (Minute 49) the Council 'decided to appoint a committee to consider what measures the Society could take to help its non-professional members.' We have already noted how the matter had arisen. The seriousness with which the Council viewed it is evident from the stature of the committee it proceeded to select: the President and his successor-to-be, two senior past Presidents, and three members who were to be future Presidents. Their report to the next Council meeting was accepted; they recommended the bimonthly publication of a 'Leaflet' containing 'signed topical and non-topical articles, and from time to time notes on matters of general interest to the body of Fellows', the first to appear in June 1938. It seemed to be understood that this Leaflets Committee would remain in being until it had launched the project. Under the title *Occasional Notes* the first issue did indeed appear in June only three months after the Council had resolved upon the undertaking — experience suggests that after nearly half a century of technological 'progress' it would now take about four times as long to achieve the same result. That issue comprised an article on 'The minor planet Hermes' by H. Spencer Jones, Astronomer Royal and President of the Society, and one on 'The quartz clock' by L. Essen. Up to 1940 March there appeared eight issues totalling 116 pages and containing 17 articles. This was in accord with 'the rate of three or four [issues] a year, in the first instance' envisaged in the President's foreword to the series.

In those days, unless he expressly desired otherwise, every Fellow automatically received *Monthly Notices* and all incidental publications of the Society, also *Memoirs* and/or *Geophysical Supplement* if he asked for them, all with no charge beyond the annual contribution of two guineas. So he received *Occasional Notes* free along with the rest and so, one assumes, did subscribers to *Monthly Notices*. Unlike other publications, *Occasional Notes* was not put on sale separately. It was a modest but interesting experiment and the Council declared itself well pleased by its reception. However, it was only just becoming established when, instead of the Society being able to work up to the projected bimonthly publication, the war brought its publication almost to a halt. To look for a moment beyond the decade, *Occasional Notes* no. 9 appeared in June 1941; it carried an Editorial Note announcing the probable cessation of publication until after the war (no. 10 did not appear until 1947 May) but announcing also the agreement with the editors of *The Observatory* whereby Fellows would thenceforth receive that magazine free of charge. The then President, S. Chapman, had already made a similar statement at the Society's meeting of 1941 April 9. To this end the Society undertook to pay a block subscription to the editors, and this in fact tided the magazine over the war years. Thus the fact of *Occasional Notes* — seen by some as a potential rival to *The Observatory* — having come into existence led to the rescue of *The Observatory* at a critical turn in its history, with manifold benefits to all concerned. To glance even further ahead, the other long-term consequence of the brief and chequered career of *Occasional Notes* was the inception of the Society's *Quarterly Journal* in 1960. Thus *Occasional Notes* may be deemed the pioneer of the 'house journal' or 'bulletin' type of publication that is now produced by a number of the world's leading scientific societies. So far as our Society is concerned, it owes this concept entirely to Dr D.H. Sadler, as I can testify from personal knowledge.

Membership

At the first meeting of the decade on 9 January 1931 the President announced that for the first time the total number of Fellows had reached four figures. This was a good start, but the attainment did not last long enough to appear in the Society's formal records. For by the end of 1931 the total had fallen to 976, and, indeed, it fell a little each year thereafter for the rest of the period.

The (Honorary) Auditor's report had always begun with a table 'Progress and present state of the Society'. This gives the statistics of membership on the last day of the preceding calendar year. The form is traditional but the grand total is a good enough measure of the numerical state. At the ends of 1930 and 1939 it stood at 973 and 901, respectively. In order to make any useful comment on the significance of any changes through the years, a fairly detailed analysis of the lists of Fellows would be needed. This would be beyond the scope of the present history. It is regrettable that no such analysis exists which could be quoted here.

Even a quite cursory inspection of just one such list does produce surprises. I chose at random that published in 1933 March. Subject to possible slips in counting, there were 916 Fellows (and no Junior Members) at that time. Of these 130 had addresses in (then) Commonwealth countries. There were 174 in other places abroad — all over the world but naturally with a preponderance in the United States. Thus out of roughly 900 Fellows, 300 came from abroad and about 600 from Great Britain and Ireland. Most of those 300 had, presumably, a fairly serious professional interest in astronomy or geophysics. Amongst the 600 'home' Fellows, I tried to count the professionals in astronomy, geophysics or related sciences and 'almost professional' amateur astronomers. I found 212; it was difficult to decide whom to include and so the precise value is not statistically significant. Most of the other 400 or so must have had an interest in knowing about astronomy, in doing some observing or simply in 'supporting' astronomy. Those were the days when popular and semi-popular books on astronomy, notably those by Jeans and by Eddington, were having unprecedented sales. So it is somewhat surprising that the number of these non-professional members did not become considerably greater. What is most surprising is that the Society could achieve so much through the exertions of a nucleus of so few active professionals.

The most immediate benefits of Fellowship to the individual were receipt of the Society's publications, attendance at its meetings and use of its library. And the main *privilege* enjoyed by a Fellow was to be able to present his work for publication by the Society, if approved by the Council. A Fellow had also the privilege of communicating papers by non-Fellows; indeed, that was the only way such papers could reach the Society. However, there was a generally accepted understanding that an author who aspired to have more than only a very occasional paper published by the Society would take steps to become a Fellow himself.

From the foundation of the Society almost to the time of World War II, it was about the only one of its sort — apart from the Royal Society with its singular status — that applied mathematicians and physicists, as well as astronomers themselves, might wish to join. In fact many did become Fellows. Indeed part of the slight decrease in numbers in the 1930s was due to the decline in that tradition. Also, of course, during much of the time concerned there was not much outlet for natural philosophers except in astronomy and related fields like optics, and others that gradually came to form the science of geophysics. Outside academic science, much of the more advanced technology of the times beside optics had to do with time-keeping, surveying, navigation and photography when it came, and these all had inevitable links with astronomy. It is surprising that the Society never did more to attract such technologists — a sign, no doubt, of the regrettable long-standing disjunction in Britain between academic science and technology.

It is perhaps a complementary characteristic that, instead of bringing in professional technologists, the Society has always attracted a considerable number of amateur scientists. During most of the period the main opportunities for amateurs were in natural history and in astronomy. It is often remarked that astronomy now remains about the only science in which amateurs can still hope to make contributions that will be recognized and used by the professionals. As noted in the first volume of the *History* (p.248) the British Astronomical Association was founded in 1890, mainly for amateurs. The happy relationship between the two societies has endured ever since but in the 1930s they were still not quite such close neighbours as they became not long afterwards, for until the spring of 1942 the Assistant Secretary of our Society occupied the flat at the top of the Society's apartments. It was only after the retirement of Miss Kathleen Williams in 1942 that the Society leased this accommodation to the BAA for its office and library. Throughout the decade — indeed from 1897 to 1942 — its meetings were held at Sion College, London EC4.

It would be pleasant to include in an account such as this something more about the personalities of Fellows who left their mark upon the Society's history in the years concerned. Space does not permit much in this way. The reader may gather some idea as to who some of these individuals were from their mention in connection with the scientific and administrative activities described. The late 1930s were the end of an era in more ways than one, for several of the most notable of these individuals died then within short intervals of one another. Indeed in 1940 November Plummer, who had been President since only the previous year, had to remark that he was for the fourth time announcing a death of one of his predecessors; F.W. Dyson died 1939 May 25, A.C.D. Crommelin 1939 September 20, R.A. Sampson 1939 November 7, A. Fowler 1940 June 24. Also one of the most devoted older Fellows H.P. Hollis died 1939 August 7, and one of the most promising younger Fellows E. Gwyn Williams 1940 May 31. Richard Inwards (1840–1937) died 1937 September 30; in 1936 February the Council had congratulated him upon the completion of 75 years as a Fellow; as a young man he had known Sir John Herschel, one of the Society's founders.

Special Occasions

In the spring of 1935 the nation and the Commonwealth celebrated the Silver Jubilee of King George V. The Society presented a Loyal Address, drawn up for the Council by the President (Reynolds), Jeans and Eddington. Also the Society contributed to the cost of decorating the front of Burlington House in honour of the celebration. The King died early in the following year and the Society presented Addresses of loyalty and condolence to King Edward VIII and to Queen Mary. At the Council meeting of 1936 May 8 the President announced that King Edward had granted his Patronage to the Society and at the next meeting that he had signed the Society's Roll. At the meeting of 1937 January 8 the President informed the Council that King George VI had, in his turn, granted his Patronage and that he was pleased to sign the Roll. An invitation for a representative of the Society to attend the forth-coming Coronation in Westminster Abbey was also announced; in due course the new President (Spencer Jones) said that his predecessor (Reynolds) would represent the Society. There were then protracted negotiations about the societies in Burlington House erecting stands in Piccadilly from which to watch the Coronation procession on 1937 May 12, but the scheme had to be abandoned. Eventually the Society was allocated ten places in one of the Government stands along Constitution Hill and these were balloted for amongst Fellows who applied.

A notable domestic occasion for the Society was a soirée in Burlington House on 1938 November 15, the bicentenary of the birth of Sir William Herschel (1738–1822), its first President from 1821 until his death. Nearly 300 Fellows and guests, including 13 direct descendants of Herschel, were present; Sir Arthur Eddington gave (twice over) a short lecture on Herschel the astronomer; Dr George Dyson, Director of the Royal College of Music, arranged for four students of the college to perform some of Herschel's unpublished musical compositions. The President (H. Spencer Jones) announced that Herschel's great grandson, the Reverend Sir John C.W. Herschel, had marked the occasion by presenting to the Society the copy by the painter Artaud of his portrait of Sir William Herschel. The copy had been painted for Herschel's sister Caroline who had taken it with her to Hannover after her brother's death. That evening it hung above the President's chair in the Meeting Room, replacing another copy (presumed not to be by Artaud himself) which in due course the Society presented to the Royal Observatory.

Occasional Notes (**1**, 27–32 and 42–43 1939) has an account of all this, but a few behind-the-scenes details may be of interest. The Council had agreed to having the celebration only on condition it cost the Society no more than £50. It regarded the first caterer's estimate of four shillings and sixpence (22½p) per head as too extravagant, and finally settled for one of two shillings and tenpence (about 14p) for buffet with claret cup, 'all buffet service, napery, cutlery, etc. included.' The Society provided sherry and the caterers provided glasses without charge.

Up to shortly before the outbreak of war, the Council had been planning another soirée to celebrate the tercentenary of the first observed transit of Venus — seen in 1639 only by Horrocks and Crabtree in England. Seemingly lest it should appear unduly eager to seize an excuse for such junketing, at the meeting of 1939 June 6 (1939 Minute 129) the Council resolved that 'a soirée should be held once in each President's period of office.' Thus it could treat the Herschel and Venus soirées as those pertaining to the presidencies of Spencer Jones (1937–39) and Plummer (1939–41). It is worth noting that, although circumstances forbade the holding of a soirée in 1939, this resolution was remembered after the war by Dr Sadler (Secretary RAS 1939–47) and soirées were in fact successfully arranged during the presidencies of H.H. Plaskett (1945–47) and W.M.H. Greaves (1947–49) but thereafter the resolution seems to have been forgotten — though never rescinded!

Numerous scientific and academic bodies invited the Society to send a representative to attend celebrations of notable occasions; amongst many others: British Association Centenary 1931; Leiden Observatory Tercentenary 1933; London University Centenary 1936; Franklin Institute dedication 1938; Royal Swedish Academy Bicentenary 1939. The Council always found some Associate, Council member or other Fellow to represent the Society, and when appropriate the Foreign Secretary prepared an address to be presented on the occasion.

Acknowledgements

Anybody now writing any of the Society's history before World War II must pay tribute to all that both the Society and he owe to those who did everything that was humanly possible to ensure the endurance of the Society and its records through the war years. The individuals concerned were few in number, without material reward, almost without recognition. They were superb in wisdom and devotion.

The writer had the advantage of some close contact with the Society during the time with which he deals but this was fairly continuous for only four of the years. Then it was almost 1980 before he had any call to write about those times. For factual details he had therefore been almost entirely dependent upon the Society's archive. At first he may have been irked by minor gaps in the record but soon the record itself showed how miraculous it is that nearly all has in fact survived.

Dr Donald Sadler played the leading part in preserving the Society's activities and its records through the war years. He has put the Society further in his debt by writing for this volume the history of those years, although as I have said this has the unfortunate consequence of rendering inadequate the account of the part he played himself. I am deeply grateful to him for reading an earlier draft of the present chapter and for suggesting how I might amplify certain items, and it was he who recalled the gift of 'presidents' received by the Society.

I am grateful to all the members of the Society's staff who have made available relevant parts of the records, particularly to Mrs Enid Lake (Librarian 1975–82) who helped so generously on the strength of her unrivalled acquaintance with the material, and to Mr Peter Hingley (Librarian since 1982) for help in later stages. I am grateful to Professor R.J. Tayler (Treasurer since 1979) for his helpfulness as Editor of this second volume of the *History*.

For assistance regarding expenses in archival researches I thank the Royal Society for a History of Science grant.

CHAPTER 3
THE DECADE 1940-50

D.H. SADLER

Introduction

With his customary facility of style Professor W.H. McCrea, in the previous Chapter, has described the general structure, organization and procedures of the Society, which had changed little over many years. In spite of the gradually improving economic situation in the country, financial support for astronomy remained static from 1930 to 1939 and the outbreak of war. This is perhaps reflected in the actual decrease in the numbers of Fellows during the period; but not, as McCrea so well shows, in the level of astronomical activity and output. There was thus little need to make more than minor changes in such well-established procedures: tradition was treasured and strongly respected — to the extent that the actual wordings of the bye-laws, originally adopted on 1831 April 6, were jealously retained whenever possible. It was a decade of stability, but not of stagnation; although relying on extremely modest resources, the Society conducted its business with quiet efficiency — and success.

McCrea's narrative account of the procedures of meetings, both of the Council and of the Society, and of the structures and activities of committees, thus provides an admirable background against which subsequent changes can be judged. This Chapter is concerned mainly with the changes arising from the disruption caused by the Second World War and the background is not, in general, described. Readers should therefore consult the previous Chapter if they wish to absorb something of the traditional dignity and stability of the pre-war conduct of the Society's affairs.

Although most of the changes were restored after the war, the later explosion of astronomical activity has made permanent changes inevitable and it may be difficult for members now to visualize — without McCrea's descriptions — how the Society functioned in the pre-war years.

(This note was added in 1983 February to a text which was otherwise substantially completed in 1977 February.)

Plate 7. Dr D.H. Sadler, President 1967–69, Secretary 1939–47. (Presidential portrait, photograph by H. Hardy FIIP, FRSA, © D.H. Sadler.)

A. The Second World War, 1939–45

1. General Observations

It is in the nature of all living organisms rapidly to recover from the stress and strain of ill-treatment, or even of actual injury and, in some cases, the organisms may emerge the stronger. Certainly there is now no evidence of the strains that affected the Society during the Second World War, and a proportion of the upsurge in the progress of astronomy, and in the fortunes of the Society, can be directly attributed to influences that arose during the war. In the history of the Society, as a continuing corporate body, the war is but an unpleasant incident, a temporary obstacle in its smooth development, and it would be inappropriate to assign to it, by a detailed historical record, an importance that it no longer possesses.

Nevertheless, the decade was dominated by the influence of the war, its effects continuing long after the cessation of hostilities in 1945, and some general account must be given of the special difficulties under which the Society continued to function. Unity requires that the story should be told from the beginning, so this Chapter takes over from the middle of 1939. However, many innovations and changes that were introduced during the war years, and perhaps because of war conditions, became firmly established; these have been treated later in this Chapter as part of the general progress and development of the Society.

2. Practical Difficulties

In 1939 the Society was particularly fortunate to have, in J.H. Reynolds, a Treasurer of great practical competence and experience. Although no-one could possibly foresee the course a war would take, or assess its effects on the Society, it was clear that war was almost inevitable and that London would, sooner or later, be attacked from the air. Reynolds, through his business connections and with characteristic efficiency, was able to arrange for the construction in the basement of Burlington House of a steel-lined bomb-proof shelter, primarily designed for the safeguarding of records; the sense of security, both for the night staff (fire-watchers) and records, that this provided proved an invaluable asset at all times. Together with H.H. Plaskett, then Secretary, he planned early the possibility of storing, at the University Observatory at Oxford, the more valuable books (particularly those in the Grove-Hills Library), equipment (for example, the Harrison regulator clock) and the irreplaceable records (signature book, lists of Fellows, minutes of Council meetings — for none of which were duplicate copies then available — as well as the Charter and the Deeds of the Society).

Special packing cases were designed and constructed and other preparation and contingency planning undertaken. At a special meeting of the Council in 1939 August 3:

> It was also decided that the eventual action to be taken would depend upon conditions during the emergency, but that as far as could be seen at present, the Society's rooms would have to be closed and the business of the Society conducted from a temporary office outside London. In this case, the Society's property left in Burlington House would be open to misuse by any temporary occupants, and it was then decided to prepare a scheme for safeguarding as far as possible the contents of the Society's rooms, such scheme to be put into action should an emergency arise.
>
> Professor Plaskett offered accommodation in the University Observatory, Oxford, for the conduct of the Society's business in an emergency. This offer was accepted provisionally, subject to Professor Plaskett receiving the approval of the appropriate authorities.

In 1939 August, many thousands of children were evacuated to the country from London and the Government had planned large-scale transfers of government offices. The decisions taken on behalf of the Society should be viewed against this background, even though they proved (fortunately) to be unnecessarily pessimistic — at least for nearly a year. They are conveniently summarized in the following memorandum to the Council, dated 1939 August 31:

> At the June meeting of Council the officers were given discretionary powers of action in the event of an emergency; this was confirmed at the special meeting of Council held on 1939 August 3 (Minute 147). These powers have been utilised, and the Society's premises at Burlington House were evacuated on 1939 August 25; the Assistant Secretary, together with all necessary records and material for the continuance of the Society's business, has moved to the University Observatory, Oxford, where accommodation has been placed at the disposal of the Society by the Trustees of the Observatory; the valuable books and manuscripts had already been sent to Oxford for safety; all remaining property of historic or intrinsic value that could conveniently be moved has been placed in the strong room constructed in the basement of Burlington House; all other moveable property, with the exception of the books in the library, has been placed in the Meeting Room, in anticipation of the premises being taken over by the Office of Works for government purposes. The Office of Works has been kept informed of the proposed plans of evacuation and of the actual steps taken; it is understood that the Office of Works will be responsible for the watching of the building during hostilities.
>
> Of the Society's staff, the Assistant Secretary has moved to Oxford where she will perform the normal duties; in consideration of her giving up her flat in Burlington House, the Officers have decided to repay the cost of her lodgings in Oxford, for the time being. Miss Wadsworth has been relieved of active duty at present, on the understanding that she should endeavour to obtain Government or other work for the duration of hostilities; in the first place the Officers propose to continue her usual salary, and

later to make up any difference there might be in the salary of her temporary post. It is understood that the porter, Steel, may join HM Forces, and it is proposed that some allowance, the exact basis of which is not yet settled, be paid to him for at least a limited period; in the meantime, while he and Mrs Steel remain at Burlington House, their joint wage will be paid in full.

During the critical period the President (H.C. Plummer) was ill in Cambridge and could only be consulted by telephone, and Professor Plaskett was in camp with the Territorial Army and communication with him was difficult; consequently the decision to evacuate and all the outstanding detailed arrangements connected with evacuation had necessarily to be in the hands of the Treasurer and the Junior Secretary.*

A meeting of Council will be called by the Officers, if further developments make it necessary.

J.H. REYNOLDS, Treasurer

D.H. SADLER, Secretary.

Upon the outbreak of war in 1939 September the President, H.C. Plummer, was still unwell, the senior Secretary, H.H. Plaskett, had already been called up, as a reservist, and was in the field with an anti-aircraft unit. D.H. Sadler, who had only taken up his full duties as Secretary in February (he had earlier deputized for W.M.H. Greaves on his translation from Greenwich to Edinburgh), was immediately transferred — with HM Nautical Almanac office — from Greenwich to Bath, the evacuation headquarters of the Admiralty. The whole burden of the immediate practical decisions arising from the war situation thus devolved on the Treasurer, J.H. Reynolds, who had already contributed so much to the preparation and planning. With the continuing help of the Assistant Secretary, Miss K. Williams, he translated those decisions into practical accomplishment, seemingly without haste or undue trouble; it was a superb example of efficient management for which the Society must always be grateful. Plate 8 is his Presidential portrait. (It was never discovered how he also found time to handle the many problems that must have arisen at his steel works in Birmingham!) The contribution from Miss Williams, both then and later, must not be overlooked; she was under considerable strain, with many difficult organizational and personal problems, and her health was not good. As reported to the Council, the office had been transferred to Oxford and the then Clerk, Miss E. Wadsworth, had been released to take up other duties; all the normal, routine, work of the Society (papers, correspondence, subscriptions, publications) had to be continued, in unfamiliar surroundings, as well as much additional work. Until the office

---

*Senior and junior normally refer to the chronological order of appointment as Secretary. At meetings the senior Secretary sits on the left of the President and the junior, on his right, normally deals with routine business. Responsibilities and duties are allocated by agreement.

could be moved back to Burlington House, Miss Williams was always overstrained — she had, for example, to transport the books, records and papers for meetings from and to Oxford — and overworked. The devotion of both Miss Williams and Miss Wadsworth to the Society and their achievements in successfully carrying through the massive reorganization that the move entailed were well worthy of the appreciation they received.

The absence of active hostilities enabled the Council to hold a special meeting on the second Friday in October (not then an appointed date of an Ordinary Meeting of the Society) in a state of relative calm. The actions already taken by the officers were confirmed, and consideration was given to a second Memorandum from the Secretary, in which an analysis was made of the possibility, or not, of the continuation of the Society's activities. The most urgent factor was that the bye-laws laid down details of dates and times of meetings, as well as of procedure; but, fortunately, it did not appear that any action likely to infringe the Charter of the Society would be necessary. The Council accordingly (extract from Minute 166):

> decided to summon a Special General Meeting of Fellows to be held immediately preceding* the next Ordinary Meeting of the Society on 1939 November 10, for the purpose of authorising any alterations to the Bye-laws that might be necessary during the national emergency, The Secretary was instructed to inform Fellows at the same time of the proposed policy of the Society and of details concerning various activities of the Society, in particular as regards the use of the Library.

It was reported that Plaskett was unable to continue his duties as Secretary and had placed his resignation in the hands of the Council; the Council deferred formal acceptance of the resignation but appointed W.M.H. Greaves and R. d'E. Atkinson to serve as Acting Secretaries. Many of the Society's activities discussed at this meeting, and the changes that were agreed and later implemented, are not further mentioned but the more important ones are referred to under their separate headings.

The Ordinary Meeting of the Society was held, in full accord with the bye-laws (except for time), on 1939 November 10; it was preceded by the Special General Meeting at which the Council and the Officers were given wide, almost unlimited powers to suspend the application of bye-laws should circumstances so require. (See under the separate heading of 'Bye-laws', p.104.) The Council had earlier held the first part of its regular meeting, which was noteworthy for the large amount of normal business (including consideration of the 28 papers received since the

---

*Necessary to allow the time of the Ordinary Meeting to be changed from 'five o'clock in the afternoon *precisely*'.

Plate 8. J.H. Reynolds, President 1935–37, Treasurer 1929–35 and 1937–46. (Presidential portrait, © Royal Astronomical Society.)

[facing page 102]

last regular meeting) that was dealt with throughout its long duration. There was, in fact, a marked return to near-normality, but with adequate provision for appropriate changes at short notice; the Secretary was able to report that the porter would continue in residence and that the rooms of the Society were to be re-opened.

The dates and times of meetings were changed, mainly because of the black-out regulations; otherwise few practical difficulties arose until the almost-continuous air attacks on London. The Treasurer reported to the Council, on 1940 November 8, that most of the windows at the rear of the Society's rooms were shattered in September and that, on October 15, the porter (R.J. Steel) and his wife had moved to the country but still attended daily. At the meeting on 1941 February 14 there was the first mention of the voluntary fire-watching and fire-fighting schemes for Burlington House; these were put into operation and continued, with the recurrent difficulties inherent in the operation of such schemes by non-residents in the area, until 1941 September when the Council decided to discontinue the voluntary scheme and to use paid fire-watchers. At that time there were about 10 volunteers from among the Fellows, together with the porter. Apart from the shattered windows (on more than one occasion), and an occasional near miss by a fire-bomb, the Society's premises had suffered no damage; but the devotion to the Society of the voluntary firewatchers cannot be too highly praised. In 1941 December the Ministry of Works and Buildings informed the Society that it would be responsible for making good any war damage; the Society's good fortune continued and no further damage was caused!

Some time later the death of T.E.R. Phillips in 1942 presented the Council with a small but tricky problem. The Reverend T.E.R. Phillips was Secretary from 1919 to 1926 and President in 1927–29. He was an accomplished amateur astronomer and had erected a well-equipped observatory in the paddock at Headley Rectory. Some of the larger instruments at his observatory were on loan from the Society and the British Astronomical Association, and clearly could not be re-allocated. An agreement was made with the new Rector to rent the paddock, the domes were camouflaged, the instruments insured and caretaking arrangements made. On 1944 September 8 it was reported to the Council that the main dome, belonging to the Society, had been damaged by a V1 flying bomb; there was no damage to the telescopes and temporary repairs were soon made.

The administrative office of the Society was moved back from Oxford to London at the end of 1942 April, and both the Assistant Secretary and the porter resumed residence on the premises. Plaskett had already generously offered to reserve the accommodation at the University Observatory should a return be necessary and to continue to

provide storage accommodation. Shortly afterwards Miss Williams resigned from the post of Assistant Secretary for health reasons, and the
Council decided to make the post non-residential, thus making available
additional rooms. Although it was many years before all the material,
including stocks of publications, that had been sent away for safe
storage was restored, the damaged windows finally repaired and some
redecoration done, this essentially marked the end of the practical
difficulties arising from the war.

3. Bye-laws

A surprisingly large proportion of the bye-laws, as they stood in 1939,
were precisely the same as were originally adopted on 1831 April 6;
there was a major addition on 1858 June 11 in regard to the procedure
for the election of the Officers and the Council but, thereafter, apart
from one or two other additions, the only changes were amendments.
There was a natural reluctance to change the long-established procedures and the bye-laws, many of them elegantly phrased, that specified them — unfortunately in embarrassing detail. Some changes were
essential (such as those specifying times, dates and places of meetings)
even on the most optimistic view of events; a realistic survey suggested
that more than half of the 86 bye-laws would need to be amended if the
Officers and the Council were to be able, as the Charter lays down, 'to
direct and manage the concerns of the said body politic and corporate'.
The following additional bye-law (No. 87) was accordingly submitted by
the Council to the Fellowship at the Special General Meeting on 1939
November 10.

> 87. Notwithstanding the present Bye-laws, the Council may, at any time
> during the present emergency when the normal procedure of calling a
> Special General Meeting is impracticable, suspend or abrogate such Bye-
> laws as may appear necessary or desirable in the interests of the Society.

The extent of the power that would thus be made available to the
Council clearly disturbed some Fellows who, in discussion at the
meeting, questioned both its validity and its desirability. It was explained: that, as far as could be ascertained, the proposal infringed
neither the text of the Charter nor its 'general scope, true intent, and
meaning'; that it was impracticable to cover, by individual amendment,
all the changes that might reasonably become necessary before another
Special General Meeting could be convened; and that the Council was
under an obligation not to utilize its extraordinary powers unnecessarily. The bye-law was adopted by a large majority.

At the same meeting two specific amendments, concerning times of meetings and remission of Annual Contributions for Fellows on Active Service with HM Armed Forces, were adopted. Fortunately, as actual events transpired, few other additional changes were required. At the Annual General Meeting on 1941 March 24 the retiring President, H.C. Plummer, summarized the actions taken by the Council; subsequently, changes in the bye-laws and all actions under Bye-law 87 were reported to Fellows in the annual *Narrative Report of the Council*. There appears to have been only one objection. At its meeting on 1941 February 14, the Council adopted a motion permitting 'wherever the Council deems it advisable and for the duration of Bye-law 87 only' election of Fellows by the Council without confirmation by Fellows at an Ordinary Meeting. This procedure was occasionally adopted when, otherwise, undue delay would have been caused. However on 1943 February 12 the Council was informed that a Fellow, who stated that he had intended to demand a ballot at the appropriate Ordinary Meeting in respect of a particular candidate, objected that he had been prevented from doing so by the Council's action in dispensing with confirmation. As the candidate had already been informed of his election and had fulfilled his obligations, the Council decided it could take no action; it is unlikely that this decision was influenced by the awareness that the Fellow (L.J. Comrie) was renowned for his strict adherence to principle and the Council's expression of regret was accepted gracefully.

Bye-law 87, together with all but two of the additional bye-laws (those in respect of the suspension of Fellowship), was repealed at a Special General Meeting on 1945 October 12. It had served its purpose adequately — in spite of its verbal and other shortcomings!

4. Meetings

Because of the black-out and travel difficulties the times of meeting were changed from the 'five o'clock in the afternoon *precisely*' of the bye-laws — to facilitate this precision a bell in the Meeting Room was actuated by direct time-signal; after some experimentation with afternoon meetings, times became standardized so that the Ordinary Meetings began at 16.30 and were timed to end at 18.00. The shorter length (1½ instead of 2 hours) matched well the reduced number of communications. The dates of meetings remained unchanged until the middle of 1940; thereafter, a pattern emerged of meeting during the lighter months (omitting January, February, November and December) with a break in August. The Annual General Meeting was gradually advanced from its traditional month of February (when it coincided with the Anniversary Meeting) to April where it remained from 1942 to 1945. The winter meetings, and the normal schedule, were restored at the end of 1945. It is, however, interesting to read, in the *Narrative Report* for 1944, that, 'The July Meeting was cancelled owing to possible danger from Flying Bombs.'

The meetings were well attended, considering the circumstances: average attendances of Fellows (as indicated by signatures in the Attendance Book, excluding visitors) rose steadily from 39 in 1941 to 56 in 1945, certainly more than half the capacity of the Meeting Room. According to precedent most meetings were taken up by the presentation of papers communicated to the Society; but the meeting in 1942 October was devoted to a celebration of the anniversaries of Galileo, Newton and Halley, and the four-hundredth anniversary of the death of Copernicus was commemorated at the meeting in 1943 May. There were no George Darwin Lectures in 1940, 1941 or 1942. One notable feature of the meetings was the diligence of successive Presidents in making it unnecessary to call upon the services of the Vice-Presidents to take the chair! They had then, of course, little opportunity for attendance at other scientific meetings or for visits abroad.

For similar reasons, attendance at the meetings of the Council — sometimes held on days other than meeting days — was uniformly high and not infrequently complete; this was quite remarkable even though ability to attend meetings was a desirable, or even necessary, qualification for nomination for election to the Council.

The Geophysical 'Secretary' (who was not one of the three Secretaries authorized in the Charter) had announced, at the meeting of the Council in 1939 October, his intention of cancelling those Geophysical Discussions that had so far been arranged; only three such meetings, one each in the years 1942, 1944 and 1945, were held before normal arrangements were resumed in 1946.

5. Publications

The decline in astronomical activity was reflected in the number and, perhaps more significantly, in the content of the papers communicated to the Society. This was, in turn, reflected in the amount of material printed in the   *Monthly Notices*. However, direct comparison of the number of pages in successive volumes is misleading. Paper rationing made it necessary to restrict, quite severely, the amount of paper used; the allocation, which varied, was almost always less than half of pre-war consumption, and at one time in 1942 was sufficient for only 300 pages of *Monthly Notices*, after allowance for a reduction in the number of copies printed. The use of a smaller type-face, increased type-area and the reduced number of copies gave approximately a 35% saving, while the actual amount of printed material was reduced by about 20–30 per cent. The reduction in the research content was, however, larger because the *Annual Report of the Council* (including obituary notices and observatory reports) and other non-scientific matter was not reduced proportionally; papers were subject to severe editing, both by their authors (at the behest of the Council) and by the editors, to remove unnecessary material and verbiage. However, there was a marked decrease in the number of papers received: from about 70 in 1939, 55 in 1940, 51 in 1941 to an average of only 36 in 1942, 1943 and 1944. The average number of pages in volumes 101, 102, 103 and 104 (covering the years 1941–44) was 330 containing about 65 per cent as many words as the 746 pages in the pre-war volume 99 (1939). Thus, although the paper situation had to be watched with care, it was never necessary to decline to print a communication for this reason alone.

The change from a winter session to a summer session provided an opportunity to associate a volume of *Monthly Notices* (e.g. Volume 101) with a calendar year (e.g. 1941); but the smaller number of pages made it impracticable to continue monthly publication, with each number associated with a particular month. In consequence of this, and the omission of the month of publication, the Council decided, in 1941 February, that the dates of receipt should be included on all papers. From volume 103 (1943) onwards, six numbers only were issued.

Few geophysical papers were received and only two numbers of the *Geophysical Supplement* were published during the war years. However, it had been originally decided to continue the series of *Occasional Notes* that had been started in 1938; five numbers (Nos. 3–7) were issued in the calendar year 1939, one (No. 8) in 1940 and one (No. 9) in 1941 — the last for several years. In that year, in view of the agreement (see p.112) with the editors of *The Observatory*, it was decided to suspend publication until after the war. Additionally in spite of all the many difficulties, a new *List of Fellows* was issued in 1943!

6. Astronomical Activity

After the outbreak of war there was, at first, only a slow decline in the number of papers communicated to the Society: 28 (then a high number) were announced as having been received at the meeting of the Council on 1939 November 10 and many of these, in addition to those referred at previous meetings, were discussed — one or two at considerable length (see p.102)! However, as more and more astronomers joined HM Armed Forces or were seconded to other duties, and many of the larger telescopes could not be fully used, the flow of papers (particularly observational papers) was reduced. A few young and some more senior astronomers were able to continue with their research. An example was the completion, in 1941, of the comprehensive discussion of the value of the solar parallax as deduced from the internationally observed parallactic displacements of the minor planet Eros, by the Astronomer Royal (Dr, later Sir, Harold Spencer Jones), for which he was, quite worthily, honoured by the award of the Gold Medal. Alas! the result, which was later shown to be affected by systematic observational errors, differed significantly from the modern value deduced from direct measures of planetary distances.

Astronomers were in great demand as scientific advisers and in the research departments. Although some stayed on after the war in their new spheres of activity, the others resumed their astronomical work with enhanced enthusiasm, and often with considerable benefit from their experiences, but generally with inadequate instrumental equipment. The void caused by the effective cessation of observation, coupled with the shortage of graduate students, was to be felt for a long time; it was compensated for by the increased general interest in, and appreciation of, astronomy (to some extent due to the black-out) and the whole new fields of research arising from the observations with the 200-inch Hale telescope at Mount Palomar and wartime research (radar, electronics, computers, atomic and nuclear physics).

The Society was mainly affected by the calls made on successive Secretaries, whose wartime duties made it impossible for them to continue in office.

Plaskett was, through the intervention of the Royal Society, transferred from his field duties to a research post with the Royal Air Force; he and Sadler were able to have occasional much-appreciated consultations when mutual air-navigational interests provided the opportunity. Atkinson, formally elected in 1940 February, gave notice of resignation in November, on appointment abroad, to be succeeded by A.D. Thackeray who was elected in 1941 March. He gave unstinting service to the Society for two years and contributed to all the many advances in that period; he would have continued but he felt that his duty lay with the Friends Ambulance Unit in the field. (After the war he resumed the secretarial duties in 1947–48.) His place was taken by H.R. Hulme, Chief Assistant at the Royal Observatory on detached duty in the Admiralty; he was able to stay until February 1946 when he decided not to return to the field of astronomy. The contributions to the war effort by Fellows of the Society were many, and would provide a notable and interesting record but they are not relevant to the history of the Society.

7. Miscellaneous Notes

(a) Medals and Awards

Mainly owing to the restriction of the freedom of choice, no Gold Medal was awarded in 1941 and 1942 and no George Darwin Lecture was given in 1940, 1941 and 1942. The election of Associates was similarly curtailed and none were elected until 1945. However the Jackson Gwilt Medal was awarded (to R.L. Waterfield) in 1942. For particulars of the awards that were made see later in the chapter (p.135).

(b) Relations with Foreign Members

The relationship with Associates and Fellows in enemy and occupied countries was discussed in the Council on many occasions. In 1942 the more general question was also raised as to the validity of electing non-British subjects as Fellows, in view of the wording of the Charter; as on previous occasions the Council (with perhaps more pressing reasons!) decided to take no action. Every legitimate effort was made to preserve contact with Associates and Fellows in foreign countries, with general rules covering suspension of Fellowship where contributions could not be sent. There was little the Society could do to assist astronomers in occupied countries but several Fellows were able, immediately after the cessation of hostilities, to make contact with them during their visits (made in other capacities) abroad and to invite them to visit the Society. It was most encouraging to meet, after such a long and painful gap, such men as J.H. Oort, who had continued to act as General Secretary of the International Astronomical Union in the most depressing circumstances; to quote the Report of the Council for the year 1945: 'Firstly, contacts with astronomers of the occupied countries of Europe have shown us that enthusiasm can overcome even the obstacles of occupation by a foreign Power.' But other qualities, not least of which was courage, were also needed. Considerable help was also given, with the co-operation of the military authorities, to the German astronomers in their struggle to resume their work. The situation returned to near-normality after the war with an amazing rapidity.

Similarly, the distribution and exchange of publications was seriously affected, but with the prior consent of the Chief Censor and the Board of Trade exchanges were continued through various channels in Sweden and the USA. In spite of paper rationing stocks of publications were reserved for distribution after the war; the subsequent restoration of library holdings in observatories and institutions throughout the world put a heavy load on library staffs.

(c) Library

Even while the administrative office of the Society was in Oxford the library was not completely out of action, as the porter was on the premises in Burlington House for most of each weekday, even when not resident. Some use was thus made of the library by Fellows, although borrowing, other than on meeting days, was difficult. From mid-1942 onwards, when the office returned to London, the usage increased rapidly as the following figures show:

Year	1938	1940	1941	1942	1943	1944	1945
Books borrowed	349	98	98	252	390	395	370
No. of borrowers	78	37	38	59	78	90	96

There is no convenient record of the number of Fellows (and others) who used the library for consultation, but it is stated to have been over 1000 annually during the later years. Certainly, the Society's premises were used to a considerable extent during the war — not necessarily for astronomical purposes, but as a friendly and peaceful haven where Fellows could relax when in, or passing through, London.

In 1942 October the Council was informed that 41 of the books bequeathed to the Society by A. Stanley Williams had 'been removed from the Society's premises' since 1939 when they had been placed in the library to await classification followed by incorporation or, if duplicates, dispersal. It was not then possible to make a complete check on the books in the library (or, in fact, on the Society's moveable property) and undoubtedly some other books were lost; but, as far as is known, nothing of any significant importance or value was lost from the Society, from any cause, during the war years. For most of this period the Assistant Secretary (Miss K. Williams, in Oxford, until mid-1942; Miss E. Wadsworth from then onwards), with the assistance of the porter, was the sole member of the staff, responsible for every aspect of administration and the library. It is a remarkable tribute to them, and to the care exercised by Fellows and visitors, that the risk of loss (to some extent deliberately accepted by the Council as the price to be paid for the continuance of facilities) was not converted into actuality.

(d) Finance

The finances of the Society were adequate for the reduced level of activity in spite of the cessation of some income from Fellows on the Suspense Account and the additional expenses caused by the war, including war damage insurance contributions. Income exceeded expenditure in each year by a small, but significant, amount, mainly as a result of the smaller (but more expensive) amount of material published and the reduction of staff salaries (with one person doing the work of two!).

In 1941 December it was pointed out to the Council that very few Fellows were available to act as Honorary Auditors, while the office was in Oxford; Madge G. Adam and D.S. Evans had performed this onerous task, which then required a detailed audit, for the previous two years and no other suitable Fellows were available. It was therefore agreed 'for this year only' to place the auditing of the accounts in the hands of a firm of professional accountants. This has been done ever since, with considerable benefit to the Society in respect of advice on accounting and book-keeping procedures and with great relief to the Honorary Auditors, who can concentrate on reporting to the Annual General Meeting on the state of the Society's premises and other non-financial matters.

(e) Lectures for the Forces

As part of the response to a demand for authoritative, but popular, lectures on astronomy the Society, in collaboration with the British Astronomical Association, organized two series of lectures for members of the British and Allied Forces, given in the Society's Meeting Room in Burlington House. The first lecture, on 1943 December 3, was given by Sir James Jeans; this was followed by five others at fortnightly intervals until 1944 February 18. A second series of six lectures was given in the following year from 1944 December 1 to 1945 February 23 but they were then discontinued. A complete list of the lecturers and their subjects is given in the *Annual Reports of the Council*, and reports are given in *The Observatory*. All were attended by large, attentive and appreciative audiences and it is to be hoped that some of those who attended continued their interest in astronomy; much gratitude is due to the distinguished lecturers who gave so freely of their time. The Society also tried to arrange, on request, suitable speakers to give lectures or talks to particular groups or units.

B. Special Items

Owing perhaps to the conditions during the war, and to the resurgence afterwards, there seem to have been, in the decade, a remarkably large number of innovations, or changes of procedure, that have since become permanent features of the Society. Many of these are due to the four outstanding Presidents of the period: S. Chapman, E.A. Milne, H.H. Plaskett and W.M.H. Greaves. Chapman, in particular, was responsible for suggesting most of the innovations during his period of presidency, as well as for firmly seeing them implemented. After the war the Society was fortunate to have, as Secretary, the supremely imaginative and competent W.H. McCrea to provide the conditions and the encouragement in which the scientific activities of the Society could begin to flourish — perhaps as never before.

The main items that have resulted in established changes are described below in roughly chronological order. A number of relatively minor items, of some interest but not leading to permanent changes, are collected together in a final paragraph.

1. The Observatory

At its meeting on 1941 February 14 the Council was informed that preliminary discussions had taken place with the Editors of *The Observatory* in regard to a proposal (made primarily because of a shortage of suitable material) that *Occasional Notes* should be merged with the magazine. The Council appointed a committee to meet the Editors; it reported at the meeting on March 14 that it:

> had reached agreement in principle on the form of 'amalgamation' to be adopted:
> (1) The Society should make an annual contribution (equivalent to a block subscription) to the funds at the disposal of the Editors, in return for which the Editors will undertake to distribute copies of *The Observatory* free to Fellows...(2)...
> (3) The scheme is to be in operation for a period of two years, with possible extension for a further period of one year.
> (4) During this period (of two years) the Society undertakes to suspend publication of *Occasional Notes*, except by prior agreement with the Editors of *The Observatory*. Such articles as would normally have been printed in *Occasional Notes* will be printed in lieu in *The Observatory*.

This report was approved by the Council, which also authorized the scheme being put into operation forthwith. The formal Agreement with the Editors, which included matters of detail, was signed on April 9; with subsequent modifications, it is still operative.

The Observatory, founded in 1877 as an independent publication, has traditionally carried reports of the scientific content of the meetings of the Society, which were not included in the formal reports of the Ordinary Meetings in *Monthly Notices*. The effect of the agreement has thus been to provide Fellows with reports, of the highest standard (of which the successive Editors can justly be proud), of the presentations and discussions at meetings which many are unable to attend — in addition to the other features of the magazine. The arrangement has been of great value to Fellows and, it is to be hoped, of considerable benefit to the Editors who were, in 1941, faced with a much reduced circulation.

The restriction on the publication of *Occasional Notes* was lifted, by agreement, to permit publication of translations of early astronomical books, but *Occasional Notes* No. 10 was not issued until 1947 May. By then there was adequate material of a suitable nature for both publications, and later it was possible to introduce the *Quarterly Journal* without conflicting with the agreement.

An interesting feature of the original agreement was 'That, while the Editors shall continue to have full financial and editorial control of the magazine, they shall include among their number a member of the Council of the Society, who shall keep the Council informed on matters concerning this agreement.' This rather one-sided arrangement was implemented, for many years, by the Council nominating one of the Editors for election to the Council when otherwise there would have been no-one in common.

2. The British Astronomical Association

The President, Chapman, suggested to the Council at its meeting on 1941 October 10 that consideration be given to possible collaboration with the British Astronomical Association, particularly in respect to accommodation for meetings. At the next meeting the Council agreed to a committee's proposal that the Society should offer, at least for the duration of the war, accommodation to the Association for meetings and for its library. After an exchange of letters between the two Presidents (the Association's President was F.J. Sellars) and a series of meetings, to discuss matters of detail, between committees appointed by the two bodies, the Councils of both agreed, in 1942 April, to the proposals: these provided for the use by the Association of the Meeting Room and the Council Room on its meeting days (the last Wednesday of the month), and for the installation of its library in the Upper Library. The latter provision meant a severe limitation on already inadequate library accommodation and could only be a temporary arrangement, made in recognition of the great difficulties that the Association would otherwise have had to face. However, before the agreement could be implemented, the unexpected resignation of Miss Williams as Assistant Secretary, in 1942 June, introduced the immediate possibility of making the post non-resident — a policy which the Officers of the Society had long recommended. The Council decided to accept the Officers' proposal to make the post non-resident; this, in turn, made available, for possible offer to the Association, much more suitable accommodation for its library in the former flat, without using the Upper Library. The offer was duly made and gleefully accepted and the formal agreement was redrafted accordingly. That agreement, with several successive modifications, still applies; the Association now has its Registered Office on the Society's premises.

The British Astronomical Association, founded in 1890, has been — and is — a valuable adjunct to the Society; relations between the

Society and the Association have been excellent, with mutual respect. Fellows can take considerable satisfaction in that the Society was able, in time of need, to give such assistance to the Association; it was much appreciated by its members at the time and continues to be a source of great satisfaction to them. Inevitably, as in any sharing of accommodation and facilities, there were a few minor difficulties but these, which were quickly overcome, were a small price to pay for the benefits to astronomy and astronomers.

3. The Narrative Report

'On the suggestion of the President it was agreed that a short account of the activities of the Society during the past year should be included in the *Annual Report of the Council...*' (1942/ Minute 59A). Thus was born the *Narrative Report of the Council*, which was first presented to the 122nd Annual General Meeting of the Society in 1942 April. The first paragraph of that *Report*, which describes its background and purpose, is given in full here:

> 1. *Introduction.* Bye-law 37 provides that 'The Council shall draw up a report on the state of the affairs of the Society, to be presented at the Annual General Meeting.' For many years the Report has consisted of a table of membership statistics, the Treasurer's Accounts, and particulars of the Society's publications, celestial photographs and instruments; the 'Report' issue of *Monthly Notices* contained also Obituary Notices of deceased Fellows and Associates, the Proceedings of Observatories and 'Council Notes on the Progress of Astronomy,' none of which can strictly be included in the Report of the Council. This plan did not give Fellows any general view of the state and activities of the Society during the Report year. To gain such a view it would have been necessary to refer to the Reports of Meetings in the *Monthly Notices*, and to examine the Council minutes, which, though open to inspection by any Fellow (Bye-law 84), can be seen only at the Society's office.
>
> The Council has therefore decided to give a narrative account of the state and activities of the Society during each calendar year; this, together with the table of membership statistics and the Treasurer's Accounts, will constitute the Council Report.'

Although its original desirability was enhanced by the rapidly moving events arising from the war, the *Narrative Report* has proved of lasting benefit to the Society as a corporate body, by providing an essential link between the Council and Fellows, most of whom are prevented by distance from attending meetings or using the Society's rooms. In addition it provides, in a convenient form, much of the material and factual detail from which the history of the Society can be compiled!

4. Accommodation for Scientific Societies

In 1944 April the President reported to the Council that the President of the Royal Society had invited the Society to be represented at a meeting to discuss future accommodation for the scientific societies; the main question for discussion was to be the possibility and practicability of a Science Centre, providing separate office and library accommodation for each society together with a series of meeting rooms, of graded capacities, and other communal facilities. After a very full discussion the Council agreed that the Society should base its attitude on a number of considerations, the first of which was:

> ...that the present accommodation in Burlington House, with the ex-
> ception of the meeting room, is likely to prove adequate for many years to
> come and that its advantages in the way of quiet, dignity and ease of access
> would be difficult to find elsewhere.

In summary, the others were to the effect that: the Society's own preference would favour remaining in Burlington House; however, it would support the establishment of a new building on a new site if this were the preferred view of the other societies; and, in that case, it would require at least the equivalent of its present accommodation.

In June the President informed the Council that the meeting had resolved:

> That a deputation, headed by the Royal Society, and consisting of one
> representative of each of the Burlington House Societies should wait on the
> Chancellor of the Exchequer, if he would receive it, for the purpose of
> presenting the urgent need for a suitable central home for science in the
> metropolis of the empire,....

Spencer Jones (Vice-President, who deputized for the President at the meeting with the Lord President of the Council, the Chancellor of the Exchequer and the Minister of Works and Planning on October 13, a meeting day) was able to report that the Ministers had been much impressed by the strength of the case for a central home for science, and had asked for a quantitative estimate of what was required. In due course a request for information on requirements was received from the Royal Society, but it was not until 1947 November that the Society was invited to appoint a representative on a committee dealing with the matter. On 1948 April 9 Sadler, the Society's representative, was invited to report verbally to the Council on the outcome of the meeting of the committee held on the previous day. It had been agreed that the committee should recommend to the Council of the Royal Society the desirability of approaching the Government to acquire an available site in West London (which had been shown to provide the floor space estimated to be required by the societies concerned) for the purpose of building thereon a 'Science Centre'. However, this scheme did not materialize, and thereafter progress was slow; but in 1950 October Sadler was able to present a confidential report to the Council in which he gave an outline of the new scheme proposed by the committee for a Science Centre on the South Bank of the River Thames, then being redeveloped. The Council agreed that full support should be given to the project, subject to later discussion on some special points. In 1950 December it was reported that the committee had been reconstituted as a Committee of the Council of the Royal Society, to include plenipotentiary representatives of the other societies to be accommodated in the Centre. The Council invited Sadler to continue to represent the Society but, for a variety of reasons which are not relevant to the history of the Society, the scheme was abandoned and the committee disbanded. It was to be nearly 20 years before more positive action was taken to improve, in particular, facilities for meetings.

5. The Isaac Newton Observatory

In his Presidential Address to the Society on 1946 February 8, Plaskett emphasized the importance of maintaining first-line observational astronomy in the United Kingdom, and advocated the erection of a large telescope of a type that would best supply the needs of British astronomers under the best obtainable observing conditions. At its meeting on March 8 the Council appointed a committee, to report not later than the May meeting:

> To consider the advisability of the Society taking the initiative, on behalf of British Astronomy, in requesting His Majesty's Government to finance the erection in this country of a telescope suitable for first-line astrophysical research; should this course be deemed desirable, to recommend to the Council the procedures for inviting the approval of the Society and for submitting the application.

On May 10 the President asked the Council, in view of developments discussed by the committee earlier that day, to approve of immediate application being made to the Council of the Royal Society for the funds necessary for the establishment and maintenance of an 'Isaac Newton Observatory'. The Astronomer Royal, Spencer Jones, explained that the urgency stemmed from the possibility that approval might be obtained in time for the announcement to be made at the Newton tercentenary celebration in July (deferred because of the war). The President read the first paragraph of the draft application, approved by the committee that morning, and said that the remainder would incorporate the committee's report on the need for, and use of, a large telescope in the United Kingdom. The Council approved that an application be sent, and it was transmitted to the Royal Society on May 14. The first paragraph reads:

To the President and Council of the Royal Society:

The Council of the Royal Astronomical Society makes application for a sum of the order of £100,000 to build a reflecting telescope of at least 72 inches aperture, together with its dome, for observational astronomy in the United Kingdom. It proposes that this instrument should be erected at Herstmonceux, the new site of the Royal Observatory, and should be under the administrative direction of the Astronomer Royal; it proposes that the observing programme should be drawn up by a Board of Management in such a way as to ensure equal opportunities for the use of the telescope to the astronomers of all the observatories in the country. In view of the forthcoming commemoration of the tercentenary of the birth of Newton and in view of the fact that the first reflecting telescope was made by him, the Society (following a suggestion of Professor S. Chapman) proposes that the instrument with its building be known as the Isaac Newton Observatory.

After a remarkably rapid and successful series of approval at all levels, the President of the Royal Society (Sir Robert Robinson) was able to announce, on July 15, that the Chancellor of the Exchequer had agreed to ask Parliament to vote a sum of money adequate to build an observatory, to be known as The Isaac Newton Observatory, equipped with a 100-inch telescope (The Isaac Newton Telescope) and its necessary ancillary equipment. The increase to 100 inches from the 72 inches of the original application can only be regarded as an added triumph for the delegation (which included the Astronomer Royal) which presented the case to the Lord President of the Council and other Ministers. A full account is given in *The Observatory* for 1946 December, and the full texts of the two memoranda submitted by the Society to the President and Council of the Royal Society were published, by permission of the Royal Society, as an appendix to the *Annual Report of the Council* for the year 1946 (*Mon. Not. R. astr. Soc.*, **107**, 11).

The Society's subsequent concern with the project was limited to the appointment of representatives on the Board of Management; periodical reports on progress were made to the Council. Unfortunately, for many and varied reasons, the speed of implementation did not match the phenomenal speed with which approval for the project was sought and obtained; the Isaac Newton Telescope was inaugurated by Her Majesty Queen Elizabeth the Second on 1967 December 1.

In this second volume of the *History* two personal notes can be added. The President was ill during most of 1946 January and February and was unable to complete the text of his presidential address. It was delivered extempore from rough notes and, to save him the task of rewriting for publication, was taken down in shorthand; those who have been privileged to hear Plaskett speak will not be surprised to learn that few alterations were necessary. The suggestion that the proposed new telescope should be linked with Newton was also made, at an early stage, by the new Secretary of the Society, McCrea; but both he and Chapman were concerned only with the happy outcome.

6. Junior Membership

In 1946 March the Secretaries placed before the Council a Memorandum on Future Policy, in which attention was drawn, in general terms, to the problems facing the Society. Arising from the discussion on this memorandum, the Finance Committee, at its meeting in December, recommended that an examination be made of methods of increasing the Fellowship; the Secretaries were asked to make suggestions for the consideration of a committee set up by the Council. This committee, under the chairmanship of Plaskett, made a number of recommendations, of which one led to the institution of a new category of membership of the Society — Junior Members. (Another outcome is described in the following sub-section.) After discussion by the Council at several meetings, at which fears were expressed and dissipated that the scheme would lead to unacceptable loss of income, the drafts of the new bye-laws required to give effect to the recommendation were finally approved by the Council on 1947 November 14. Their adoption by the Society was proposed by the Council at a Special General Meeting, called for that purpose on December 12, and was duly agreed.

Essentially, within specified age limits, Junior Members are able to enjoy almost all the practical privileges of Fellowship for a much reduced annual contribution; the main restriction is that, not being Fellows, they have no voting rights and cannot hold office. In its *Annual Report* the Council expressed the hope that the institution of Junior Membership would encourage those who would normally join the Society to do so at an earlier age, and would also attract others who might not otherwise join; there can be no question that its hope has been amply fulfilled.

7. Meetings outside London

The *Annual Report of the Council* for 1946 contained the sentence 'Although relatively few Fellows are able regularly to attend meetings of the Society, they are the very essence of a corporate body and it is the Council's earnest endeavour to encourage Fellows to attend whenever possible.' One of the recommendations of the Plaskett committee (see the preceding sub-section), approved in due course by the Council, was that meetings of the Society should be arranged at other centres of population in the United Kingdom; these meetings were to be meetings of the Society in the full sense of the bye-laws and not merely meetings held under its auspices. The first such meeting, necessarily something of an experiment, was held on 1948 October 29 in Edinburgh, and was a resounding success (see Plate 10). This success was due, in very large measure, to the enthusiastic support and assistance (as well as hospitality!) given by many scientific institutions and civil authorities in Edinburgh, as well as to the superb planning and organization by the President (W.M.H. Greaves, Astronomer Royal for Scotland). The Council, very gratified with the experiment, agreed that a similar additional meeting be held during each calendar year. An equally successful meeting, with equally enthusiastic support, was duly held in Manchester on 1949 July 1 with the co-operation of the University of Manchester and other organizations; it provided the occasion for the first visit, for Fellows of the Society as such, to the Jodrell Bank Experimental Station, 'where much of the pioneer work in radio astronomy is being conducted.' Both these meetings were held with the full co-operation and participation of the local astronomical societies and, apart from the associated lectures, visits and receptions, served well their intended purpose of enabling many astronomers, whether Fellows or not, to attend an Ordinary Meeting of the Society.

Further details of the arrangements for these meetings, of the functions, visits and lectures held in association with them, and of the Council's expression of the Society's appreciation of, and gratitude to, the many who contributed to their success are given in the *Narrative Reports*.

8. The Eddington Medal

In 1946 June the Secretaries proposed to the Council that the Society should commemorate, in some suitable way, the outstanding services of Sir Arthur Eddington OM to Astronomy and to the Society (he was Secretary in 1912–1917, President in 1921–23, Foreign Secretary in 1936–44 and a most assiduous member of the Council for many years). They further suggested that this should take the form of an 'Arthur Stanley Eddington Prize' to be awarded at intervals for the most outstanding contribution to *Monthly Notices*. This suggestion was favourably received and the Secretaries were asked to formulate it in more detail; but when their draft regulations were eventually discussed, in 1947 April the Council decided not to proceed with the plan. In lieu it was proposed that the Council should sponsor an appeal to raise a fund (The Arthur Stanley Eddington Commemoration Fund), the income from which could be used, in part, to finance a medal, prize or lecture.

In due course, in 1947 November, the Council approved a committee report proposing the institution of the Eddington Gold Medal:

> for outstanding work in theoretical astronomy, more especially in those branches to which Eddington contributed but not excluding celestial mechanics

and the launching of a direct appeal to Fellows for contributions to the fund. An announcement to this effect was made by the President at the Ordinary Meeting on that day. However, the appeal did not go out until 1948 April and did not, at first, meet with the response that was hoped for and expected; it was not until early in 1950 that the Council's approval was advanced to the stage of agreeing to draft regulations for the award. These were incorporated into the new bye-laws that were adopted by the Society at the Annual General Meeting of 1951 February 9.

The Council was aware of possible conflict, or duplication, between the grounds of the award and the more general grounds for the award of the Gold Medal — the Society's highest award, dating from the beginnings of the Society itself. It endeavoured to ensure that, in making an award, a clear distinction be made between the two.

9. Procedure at Meetings

Traditionally there are no Standing Orders, as such, for the conduct of the meetings of the Council, except in respect of the procedures for the award of the Gold Medal and the selection of the Council's nominations for inclusion on the Balloting List for the next Council. In response to a request, in 1941 June, the Secretaries were asked to draw up a list of procedural rules to incorporate those specified in the bye-laws and arising from earlier Council decisions. The Council later accepted the Secretaries' proposal to prepare comprehensive 'Notes for the Guidance of Members of the Council'; these were approved in 1942 March. They included explanatory information as well as the relevant provisions of the bye-laws, such rules of procedure as existed, previous decisions of the Council (for guidance and not necessarily binding) and a monthly schedule of the recurring items of agenda. It was assumed that the President (or other Fellow in the chair) would conduct the meeting in accord with usual committee practice and with his duty as specified in the bye-laws:

> ...to regulate and keep order in all their proceedings; to state and put questions according to the sense and intention of the meeting; and to carry into effect the Bye-laws of the Society.

A second edition, necessary to incorporate the return to normal conditions, was prepared in 1946 March.

Much of the business discussed at meetings of the Council was routine (including details of bank balances, minor financial transactions, payment of bills in excess of a small limit, the reading of lists of candidates for proposal and election) and this was increasingly blamed, often unjustifiably, for over-running the normal time for rising — thus curtailing the time available for tea and, more important, meeting other Fellows. But it was not until 1950 May that the Secretaries proposed, and the Council accepted, a scheme in which reports on routine matters (including lists of candidates) were circulated with the agenda and accepted, without discussion, unless queried.

Similar criticism was directed at the practice, enjoined by the bye-laws, of routine business at the Ordinary Meetings; on many occasions, in and out of the Council, the relative merits of traditional procedure with treasured formulae (conveniently placed to aid presidential memories!), on the one hand, and a few minutes of additional time for scientific discussion, on the other hand, were argued. Expect on a few special occasions, when unreasonable delay would otherwise have been caused, the Council decided to preserve the tradition, even when it could have invoked Bye-law 87 to omit or shorten it. Fortunately, since the demand for a ballot refers to all candidates and not to a particular one, the Society was not often asked to ballot for the election of Fellows. The Society possesses some 20 ballot boxes and an awkward situation could have arisen if there had ever been more than 20 candidates for confirmation of election by the Council; furthermore, there seems to be no definite rule as to the majority required for election. At the meeting on 1946 June 14 when a ballot (for which advance notice had courteously been given to the Secretaries) was demanded there were only 7 candidates — all were confirmed.*

According to the bye-laws the main business of the Ordinary Meetings was 'to read such communications relating to Astronomy and the subjects connected therewith, as may have been presented to the Society'; it was the tradition, and a secretarial duty, to 'read' papers even if the author could not do so personally (though most were read by title only). In 1942 the Secretaries proposed that organized discussions should be arranged instead, to which the Council somewhat reluctantly agreed 'provided there was insufficient papers for reading.' One or two such discussions were, in fact, held during the war years.

A postal ballot was held in the summer of 1945 to seek the views of Fellows on the time at which Ordinary Meetings should commence and whether there should be a meeting in June or October. As a result the Council agreed to recommend that the bye-laws be changed so that

*This could be the last occasion on which a ballot for confimation of election was held, since the bye-laws were changed in 1977 March to eliminate the confirmation, at an Ordinary Meeting, of the election of Fellows by the Council.

Ordinary Meetings be held in each month of the Session from October
to May, commencing at 16.30 ('half-past four o'clock in the afternoon
precisely'). This change was adopted by the Society at the Annual
General Meeting on 1946 February 8. Another small, but significant,
change was also made to differentiate more clearly between the two
meetings normally held on the second Friday in February — the
Anniversary Meeting (which is an Ordinary Meeting) and the Annual
General Meeting.

10. Miscellaneous

(a) A. Stanley Williams' Bequest

In 1938 A. Stanley Williams, a skilled and indefatigable observer of
variable stars (for which work the Society had awarded him the
Jackson Gwilt Medal in 1923), bequeathed to the Society (in addition to
his books and a 9.5-inch telescope) his voluminous manuscripts —
containing about 10000 observations of variable stars, many of them
unreduced — together with a sum of money partly to be used for
reduction and publication. The Council appointed a committee to make
recommendations, as a result of which P.M. Ryves (another experienced observer of variable stars) was invited to make a detailed
examination of the whole material, which he duly did. However, it
proved impossible, during the war, even to carry out the trial reductions
which were an essential preliminary to treatment of the whole material;
the project dragged on, with innumerable discussions and much devoted effort (particularly on the part of Ryves), until the committee
prepared a long memorandum dated 1949 October 9. The proposals
therein were accepted by the Council early in 1950 and led, eventually,
to partial publication, some material in full but the less important in
summary only. The total effort, as the Secretaries can testify, seemed
quite disproportionate to the final result; the Society is not well fitted to
undertake such work.

(b) The Better Sky

In 1942 December the President reported that, the well-known author
A.P. (later Sir Alan) Herbert had written to suggest that the constel-
lations should be renamed to accord more appropriately with modern
views and, in particular, navigational applications. Although interested
in and sympathetic to the suggestion, bearing in mind the greatly
increased use of astronomical navigation during the war, the Council
could clearly take no action (even if it had wished to do so) on a matter

that would require international agreement. The detailed proposals were later published under the name *The Better Sky.*

(c) Gill's Heliometer

The Secretary reminded the Council, in 1943 February, that a predecessor (H.H. Turner) had planned to have a brass plate affixed to the table in the Meeting Room to explain the depression caused by the fall of Gill's (later Sir David Gill) heliometer in 1877. The Council agreed that a minute should be recorded so that the matter should not be overlooked when it could be implemented after the war — but the table, without the plate, is no longer in the Meeting Room.

(d) The Discovery of Neptune

The Council invited W.M. Smart (one-time John Couch Adams astronomer, and known to have much material including a draft biography relating to Adams) to give an address on 'John Couch Adams and the Discovery of Neptune' at the conversazione on 1946 October 8 to mark the centenary of the discovery. This he duly did (in fact, owing to the limited capacity of the Meeting Room, he gave it twice!) and it was later agreed to print the full text (of which the address was a much shortened summary) in *Occasional Notes* (No.11, 1947 August). The Astronomer Royal (Spencer Jones), who was Treasurer at the time, at first objected that it was too long but the Council, at its meeting in 1946 December, decided not to ask that it be shortened; at the proof stage he then asked that certain passages (relating to Adams' visit to G.B. Airy at Greenwich in 1845 October) should be amended. The author defended his version of this rather controversial episode and the editor declined to intervene. The Astronomer Royal later (after publication) gave his version of this visit.

(e) Observatory House, Slough

In 1946 August the Reverend Sir John (C.W.) Herschel Bt. wrote to the President offering the Society a long-term lease of 'Observatory House' at Slough, in which certain instruments and articles of historical interest would be left. He suggested that the house might be used as an extension of the Society's premises, possibly for accommodation of visiting astronomers. After a very full discussion the Council decided that, with much regret, the Society could not possibly consider accepting the offer owing to the heavy financial commitment it would entail and the small use to which it could put the house. The house has, inevitably, since been sold (and subsequently demolished) and the contents dispersed.

(f) Astronomical Navigation

Until recent years the Society played a large part in the design of *The Nautical Almanac*. Its direct interest dates from 1830 when The Lords Commissioners of the Admiralty requested its advice; the comprehensive reforms that it suggested were introduced in the edition for 1834. In 1890 the Society took the initiative in suggesting changes to the Admiralty and these were duly adopted in the edition for 1896. Furthermore, in 1897, a dignified protest that it had not been consulted (about the new values of the fundamental constants agreed to in 1896) brought forth an apology from the Admiralty. The Society was again formally consulted in 1910 and its recommendation, and detailed design, for a separate publication for navigation at sea was implemented in 1914. The last occasion of formal consultation appears to have been in 1917 when the Society was asked to express its views on the proposed change of the beginning of the astronomical day from noon to midnight. Possibly by oversight, it was not consulted about the changes introduced in the navigational almanac in 1929 nor, rather more particularly, about the comprehensive changes in *The Nautical Almanac* itself in 1931.

It was therefore not surprising that the function of advising the Admiralty (and later other bodies concerned with astronomical navigation) should pass, in 1947, to the newly formed Institute of Navigation (now The Royal Institute of Navigation); the Institute was invited in 1948 to advise on the redesign of the almanac for surface navigation, the new form being introduced for the year 1952. Thus came to an end an interesting phase in which the Society, which especially in the later years was by no means representative of navigational practice at sea, gave sound advice and acquitted itself with considerable credit.

C. Continuing Themes

The division into decades introduces either discontinuities or overlaps into themes that are continuous. Since the war years, 1939-45, have also been treated separately there is the further danger of repetition within the decade.

1. The Society's Officers

As mentioned earlier, the Society was fortunate to have a succession of outstanding Presidents to direct its affairs during the decade. When H.C. Plummer died in 1946 the Council was able to say: 'That the Society has weathered so well the shock and strains occasioned by the war is due in no small measure to Plummer's wisdom and guidance in those first critical years. His old-world courtesy, his modesty and quiet manner provided an atmosphere of stability and peace that was so missing elsewhere.' Plummer had been Professor of Astronomy in the University of Dublin and Royal Astronomer of Ireland. He was an authority on celestial mechanics and his book on *Dynamical Astronomy* (1918) was the standard work. He devoted the 1940 Presidential Address to an account of the work of Edwin P. Hubble, the Gold Medallist, and his second address, in 1941, to the 'Development of the Vertical Telescope.'

In some contrast S. Chapman brought great qualities of vision and decision; an outstanding scientist, he was also a most competent manager who firmly translated his many ideas into actions — many of the innovations mentioned in the previous section owe their inception, and realization, to him. Although he could be classed equally well as an astronomer, Chapman was renowned as a geophysicist and can be regarded as the first academic 'geophysical' President of the century. It would be invidious to pick out any of his work for special mention, though his Presidential Addresses give some indication of the breadth of his interests. His 1942 address was on 'The Source of the Sun's Energy', which was prophetic; and in 1943 his address was devoted to 'Magnetism in the Sun's Atmosphere.' His contribution to *Occasional Notes* (No. 8 on *Edmund Halley as Physical Geographer and the Story of his Charts*) did a great deal to set a standard of excellence, which unfortunately occurred at a time when it was the last number to be issued for several years. Plate 9 is his Presidential portrait.

He was succeeded by E.A. Milne who, on the other hand, combined a certain amount of caution with sound judgement, extreme conscientiousness and a completely objective outlook. He was currently Rouse Ball Professor of Mathematics at Oxford. His two Presidential Addresses indicated his chief interests in astronomy: in 1944 'On the Nature of Universal Gravitation' and in 1945 on 'The Natural Philosophy of Stellar Structure.' He was the originator of kinematic relativity and defended his views with passionate argument. His debates, with Eddington, on both these subjects were carried out at meetings with mutual respect and, in the case of Milne, with considerable eloquence. Plummer, Chapman and Milne were, in their different ways, admirable Presidents, although none had previous experience of the Society's affairs as a Secretary.

They were followed, in turn, by H.H. Plaskett, W.M.H. Greaves and W.M. Smart who had all been Secretaries (though, curiously enough, in the reverse order to that in which they became President) and who were admirably qualified by temperament, training and experience to guide the Society to stability and to achievement through the turbulence left by the war; each made significant, and lasting, contributions to the well-being and standing of the Society.

At one time it was the custom for a President to devote his annual Presidential Address to a full-length statement of the grounds for the award of the Gold Medal and an account of the medallist's work — a formidable task if their fields of interest were far apart. This tradition was followed in 1940 (Plummer on Hubble), 1949 (Greaves on Chapman) and 1950 (Smart on Stebbins), but otherwise the Presidents spoke on subjects of their own choice.

Plaskett was Professor of Astronomy at Oxford and had played a large part in providing accommodation for the Society's records during the war in Oxford. References (see p.116) have already been made to the impact of his Presidential Address in 1946; his 1947 address was on his own particular subject 'Astronomical Spectroscopy.'

Greaves was a former Chief Assistant at the Royal Observatory, who became in 1938 Astronomer Royal for Scotland after the retirement of R.A. Sampson. He was a man of sound judgement and distinct personality, and treated all matters concerning the Society with conscientiousness. His 1948 Presidential Address was concerned with his own subject, developed at the Royal Observatory, of 'The Photometry of the Continuous Spectrum'; this covered the concept of colour temperatures and the corresponding gradients.

Smart, once John Couch Adams Astronomer at Cambridge, was Regius Professor of Astronomy at Glasgow. He will be remembered by many for his style of writing textbooks; his book on *Spherical Astronomy* was a classic.

Reynolds' contribution to the wartime running of the Society has already been mentioned and it must be repeated that it was enormous. He was also an outstanding Treasurer, with sound financial experience and excellent business sense; the Council sought, and benefited from, his special advice not only on staff matters (salaries, contracts of service, pensions, insurance, etc.), but also as regards the upkeep of the premises, instruments and, essentially, all business and practical matters. He retired as Treasurer, under the Council's ten-year rule, in 1946. When he died, at the end of 1949, the Council could record its appreciation (after referring earlier to his scientific work and to his munificent gifts to astronomy) in the words:

> But perhaps his greatest contribution to our science lies in the selfless devotion with which he advanced the aims of the Society by willing service as Treasurer, as Vice-President and as President. On his retirement from office in 1946 he had completed a service totalling more than a quarter of a century, in one capacity or another, on the Council of the Society.

He was succeeded as Treasurer by Spencer Jones, former Secretary

Plate 9. Professor S. Chapman FRS, President 1941–43, Gold Medallist 1949. (Presidential portrait, © Royal Astronomical Society.)

[*facing page* 126]

and President, who conducted the financial affairs of the Society with his usual quiet efficiency.

At the outbreak of war the office of Foreign Secretary was filled by Eddington, then President of the International Astronomical Union and one of the world's greatest scientists. Although a few problems (refugees, foreign Associates and Fellows, exchange of publications) arose during the early days of the war, there was little that could be done in respect to foreign relations. His death, at an early age, at the end of 1944 robbed him of the pleasure that he would certainly have had in welcoming, only a year later, the astronomers from war-torn Europe. It also robbed astronomy of a unique intellect and the Society of one of its staunchest supporters; also a former Secretary and President, he served on the Council for many years, rarely missing a meeting and always ready to give of his time.

Eddington was succeeded by F.J.M. Stratton, so well fitted by temperament and experience to represent the Society in its relations with foreign astronomers and institutions. He was currently Secretary General of the International Council of Scientific Unions and had been General Secretary of the International Astronomical Union — an ideal person to renew the collaborations and friendships disrupted by the war.

With the exception of Sadler, none of the wartime Secretaries (Atkinson, Thackeray or Hulme) was able to continue in office for more than a few years and then only under grave difficulties. In particular Atkinson and Hulme both carried not only a heavy work-load but also high responsibility for operational decisions. Hulme used to say that the Society's problems, which he tackled with calm precision, provided him with essential relaxation! The Society owes much to Thackeray's quiet efficiency and supreme conscientiousness, and to them all for their solid contributions, their enthusiastic ideas and, above all, for their so-valuable time; Sadler, with whom they worked, acknowledges their willing co-operation with him and the help and encouragement that they gave.

McCrea, being then available to attend meetings, was elected to the Council in April 1944 and in due course succeeded Hulme as Secretary in 1946. In the three years that his other duties allowed him to remain in office he was largely responsible for transforming the Society in anticipation of the upsurge of astronomical activity that was to follow. Most of the far-reaching changes in this period owe their origin, and their development, to him. He was much helped by the unselfishness of Thackeray who returned to office in succession to Sadler in 1947; although technically senior by way of appointment he (i.e. Thackeray) insisted on taking the seat on the right of the President — as 'junior' Secretary (see footnote p.101.) However, he was only able to serve for little more than a year before resigning on taking up the post of Chief Assistant at the Radcliffe Observatory in Pretoria; the Council, in accordance with Bye-law 7, appointed R.O. Redman in his place from 1948 June.

At the Annual General Meeting in 1949 February, A. Hunter was elected Secretary in place of McCrea, who had served the Society so well, and Redman's appointment was confirmed. However, in 1949 May Redman requested permission to resign owing to the pressure of his duties as Director of the Cambridge Observatories. The Council, again invoking Bye-law 7, appointed Flora M. McBain (later Mrs D.H. Sadler) to succeed him as from 1949 October; she was the first woman to become a Secretary of the Society.* At the same time she was relieved of her editorial duties by the appointment of G.J. Whitrow as Editor and he was invited to attend Council meetings until 1950 February — in anticipation of being elected to Council at the Annual General Meeting (as he was). There were thus a number of changes, in rapid succession, at a time when the resurgence of astronomical activity was placing increasing pressure on the Society. Fortunately, Hunter and Miss McBain were able to serve for several years together.

The Charter allows the Society to have 'not more than three Secretaries', so that for many years a member of the Council (who was excused from the normal retirement rules) was appointed to act as 'Geophysical Secretary', without formally being a Secretary of the Society. R. Stoneley served in this capacity until 1946 when he was succeeded by B.C. Browne. Although there were few meetings, few papers and little geophysical activity during the war, Stoneley's services to the Society as a member of the Council for many years were considerable. Browne was able gradually to restore the normal programme of Geophysical Discussions and publication of the *Geophysical Supplement*. Interest in geophysics was slower to reflect the post-war upsurge than astronomy but was equally dramatic in intensity when it did come.

2. Membership and Finance

There was little change in the number of Fellows during the war years: 857 on 1939 January 1, 777+89 (non-active, on the suspense account) on 1946 January 1. This number rose to 893+13 on 1949 January 1, but there was a marked rise in total numbers in 1949 to 993+10 (non-active)

---

*Miss McBain was not the first woman to serve on Council. That honour belongs to Madge Adam who was elected to the Council in 1943, thus ending a period of over 27 years during which women had been eligible for Fellowship and service on the Council and none had served. Subsequently, as a Vice-President, she was the first woman to chair a meeting of the Society.

+30 (Junior Members). The number of Associates remained almost constant at rather less than 50; it was not until later, in 1950, that the Council revised the earlier rule of a maximum of 50 to 'not more than three elections in a year'.

On the other hand there was a marked increase in the annual income and expenditure, from about £3250 in 1939 (decreasing to about £2800 in 1945) to an average of £5500 in 1948 and 1949. Owing primarily to reduced staff and smaller publications there was a small surplus in most war years but a large compensating deficit in 1946. After 1946 small annual grants were received from the Government Grant-in-Aid for Scientific Publications, thus reducing what would otherwise have been large deficits, as wartime arrears of printing were overtaken. The Annual Contribution was increased from 2 guineas to 3 guineas at the Annual General Meeting on 1947 February 14, the first increase in the Society's history!

An interesting situation arose in connection with the corresponding increase in the composition fee by which Fellows could compound, either at election or later, for future annual contributions. Fellows must be given notice of any proposed change in the bye-laws, and such change only becomes effective when approved by a General Meeting; thus there will be a period during which Fellows can compound at the old rate. The Council decided, deliberately, not to introduce any restriction, voluntary or otherwise, on composition during this period, and nine Fellows availed themselves of the opportunity to compound, at the old rates, before 1945 October 12, on which date a Special General Meeting approved the increase proposed by the Council, in anticipation of the later increase in the annual contribution. The Council then issued an invitation to *all* compounders to make an additional contribution, making it clear, however, that there was no actual or moral obligation; several years later, it firmly rejected a proposal to prohibit compounding between the announcement of a proposed increase and its consideration at a General Meeting.

In 1949 February the Council gave consideration to a 'Memorandum on the desirability of securing a grant from Public funds' prepared by the Secretary (McCrea). This arose from the fact that the Society, in spite of economical management, had an annual deficit of £1000 despite a 50 per cent increase in subscription; this was at present paid out of a Parliamentary Grant-in-Aid. The Council approved the paper but appointed a senior committee to explore the matter thoroughly. The full report, addressed to the President and Council of the Royal Society, was presented to the Council in April but discussion was deferred until the Treasurer could be present. In May the Council approved the report, with some amendments and suggested that the Treasurer should discuss it, unofficially, with the Secretaries of the Royal Society. The Memorandum was duly handed (officially) to the Secretary of the Royal

Society in 1950 February, when the Treasurer 'gave to the Council a confidential report of the Royal Society's views.'

The substance of the Memorandum was, in the case of the Society, that it required an annual grant (of the order of £5000) from public funds sufficient to ensure that it could achieve its objective, but that the Council believed that other learned societies may be meeting difficulties similar to their own and that the Royal Society should consult other societies preferably before approaching HM Government.

While it is not relevant to the decade under consideration, the Memorandum gave rise to the setting up of a Royal Society committee on the publications of the various societies, with an expert advisor mainly to increase sales to the public. This was the earliest beginning of the substantial change in the Society's policy regarding publications.

During the decade, including 1939, the Society lost through death 15 Associates (statistically a small number), its only Honorary Member (Annie J. Cannon) and no fewer than 10 former Presidents. In chronological order of their deaths the Presidents were:

F.W. Dyson 1911–13; also Treasurer 1935–37 and Secretary 1899– 1905;

A.C.D. Crommelin 1929–31; also Secretary 1917–23;

R.A. Sampson 1915–17;

A. Fowler 1919–21; also Secretary 1913–19 and Foreign Secretary 1931–36;

T.E.R. Phillips 1927–29; also Secretary 1919–26;

A.S. Eddington 1921–23; also Secretary 1912–17 and Foreign Secretary 1936–44.

H.F. Newell 1907–09; also Secretary 1897–1901;

J.H. Jeans 1925–27;

H.C. Plummer 1939–41;

J.H. Reynolds 1935–37; also Treasurer 1929–35, 1937–46.

In the annual lists of deceased Fellows there occur the names of many famous people, some of whom (such as R.H. Fowler and G.H. Hardy) were not primarily astronomers, including A.R. Hinks who was Secretary in 1909–13. Obituary notices are given for all of the above in the *Annual Reports of the Council*, together with those of many other Fellows. Until 1946 it was the custom to include notices for all deceased Fellows, even though it was often difficult to obtain adequate information, but in 1946 the Council decided that obituary notices would, in future, be given only for those who had made some significant contribution to astronomy (or geophysics).

It is not possible adequately to pay tribute here to the contributions made to astronomy, and to the Society, by these former Presidents, but special mention must be made of three (in addition to Eddington,

Plummer and Reynolds already mentioned elsewhere). Sir Frank Dyson, ninth Astronomer Royal 1910–33, was President of the International Astronomical Union (IAU) in 1928–32 and a much respected, and loved, man. Alfred Fowler, the first General Secretary of the IAU in 1919–25, had served the Society well since 1889 and was a Vice-President when he died, having been so elected in 1940 February; his wisdom and experience would have been of great value. Sir James Jeans OM served on the Council in 1940–42 and thereby added to the service he had already given to the Society. As the Council resolution said on 1946 October 11: 'He was besides [his scientific work] perhaps the greatest of expositors of modern physical sciences...a presidency signalized not only by three brilliant addresses on the awards of the Gold Medal, but also by his endowment of the George Darwin Lectureship.'

Two bequests deserve special mention. H.C. Plummer bequeathed the sum of £1000 to the Society and made it one of his residuary legatees; a total sum of £1500 was eventually received, with very broad conditions on its use. D. Carder-Davies (elected a Fellow in 1923) bequeathed, free of duty, the sum of £3000 with the following stipulation, '...that the legacy should be without condition, but he asked that no part be spent on any purpose controlled or suggested by the Government.'

3. Bye-laws

Several alterations to the bye-laws, mainly of a minor character, were proposed by the Council at the Annual General Meeting on 1946 February 8 and adopted. There was one important alteration of principle, namely the new Bye-law 48 which prohibited the Council from bringing forward, at the Annual General Meeting, business of which prior notice had not been given to Fellows; the Council, having enjoyed and invoked much greater freedom during the war years, itself proposed this limitation on its own powers! Other changes concerned the months and times of Ordinary Meetings and the distinction between the Anniversary and Annual General Meetings in February. The Session was to be from October to May and the time of meeting 16.30 (see p.122). An alteration to the scale of composition fees had previously been made at the Special General Meeting on 1945 October 12, and effect was given to the related increase in the Annual Contribution at the Annual General Meeting on 1947 February 14; at this same meeting a new section of the bye-laws was adopted to provide for the introduction of Junior Membership. The various temporary bye-laws, relating to wartime conditions, had already been repealed. Allowing for these few major changes, and a number of minor verbal clarifications, the bye-laws at the end of the decade were thus essentially unchanged from those at the beginning and most preserved their original wording. In the new edition of the bye-laws, the opportunity was thus taken of attaching to each bye-law the dates of its original adoption and of its most recent amendment. Procedure at the Ordinary Meetings was still rigidly defined and formalized, and the dates and times of meeting were inflexible. The return to such tradition was, to most Fellows, a source of satisfaction and an expression of confidence in future stability.

One of the more controversial aspects of the Society's procedure is the method of nominating and electing the Council. For many years Bye-law 8 (originally adopted on 1858 June 11), allowing nominations to be made from outside the Council, had not been invoked, thus leading to a balloting list containing only the complete list of the Council's nominations and to a ballot that was a formality. In spite of this apparently undemocratic procedure, the Council's list was selected from many nominations (from within the Council), and based on a large amount of information and many considerations not available to Fellows as a whole; the selection process was governed by elaborate rules of procedure and it almost certainly resulted in a more representative Council than a contested election, in which many votes would necessarily be cast more-or-less at random. Fully aware of this two former officers (Milne and Sadler) nominated, at the end of 1948 in accordance with Bye-law 8, a Fellow (G.J. Whitrow) for membership of the Council: first, because they considered that he would be a very useful member of the Council (as he, in fact, later became), but also to draw the attention of Fellows to the existence of their right to nominate. There was no collusion with the Council but the purpose of the nomination was made clear. It might reasonably be expected that a candidate, so nominated, would stand little chance of being elected but, at the Annual General Meeting on 1949 February 11, Whitrow received the same number of votes as one of the Council's nominations (P.J. Treanor); the President (W.M.H. Greaves) gave his rarely used casting vote in favour of the Council nominee.

4. Meetings

With the return to near-normal conditions in 1946, it was possible thereafter to revert to the pre-war schedule of meetings, with a change of month (October instead of June) and time. In 1946, however, a special meeting was held in January in honour of several French astronomers then visiting the country, and an additional meeting was held in June. Additional meetings were also held outside London in both 1948 (Edinburgh) and 1949 (Manchester). Plate 10 is a group photograph from the Edinburgh meeting. Several meetings were devoted to

Plate 10. The Society's Meeting at Edinburgh 1948 October 29-30. Group photograph at the Royal Observatory Edinburgh. The President, Professor W.M.H. Greaves, who was also Astronomer Royal for Scotland. is in the centre of the front row. (© The Scotsman Publications.)

organized discussions, partly still owing to the shortage of papers suitable for presentation at meetings. One meeting was usually largely taken up by the George Darwin Lecture and the February meeting by the President's Address. Average attendances increased gradually.

Although it had been possible only to arrange one Geophysical Discussion in 1945, a full programme of five meetings was held in 1946, increased to six in each of the years 1947, 1948 and 1949 (with an additional meeting in Manchester). Attendances were encouraging and the audiences (which included many non-Fellows) were appreciative. The range of subjects was wide: it is interesting to note that there was a discussion on 'English Oilfields' on 1946 October 22, and an exhibition of the then new quartz-crystal oscillators on 1947 March 21 in connection with the discussion on 'Modern Methods of Time-keeping.'

It had been agreed by the Council in 1939 that the Society should hold an evening reception, or conversazione, once during each President's term of office, and that arrangements should be made for a Christmas lecture for children or young people. Neither could be implemented during the war and the suggestion of a Christmas lecture was not revived; however, some celebratory functions were held.

The first, to which reference has already been made, was held on the evening of 1946 October 8 to mark the centenary of the discovery of Neptune; the opportunity was also taken to commemorate the quatercentenary and tercentenary, respectively, of the births of Tycho Brahe and Flamsteed. After the austerity (which still, in fact, continued) of the war, the conversazione was well attended and proved to be an outstanding success.

The second event was the reception and conversazione on 1948 July 23, arranged as late in the season as practicable to enable the Society to entertain astronomers passing through England on their way to attend the first post-war General Assembly of the International Astronomical Union in Zürich. Fellows and visitors were favourably impressed by the many exhibits and demonstrations that had been organized.

Social events were also organized, but not by the Society, at the two out-of-London meetings; these were much appreciated and contributed markedly to their success.

5. Publications

The resumption of scientific activity after the war coincided with a grave shortage of the skilled compositors who typeset mathematical material. The output of the Society's printer (who had served the Society long and well) fell alarmingly, and it became necessary to search for a new printer with suitable experience, equipment, and staff — and then to negotiate a new contract and agree on a new style (the opportunity was being taken of returning to pre-war standards of presentation). As a result very little material could be printed in 1946 and large arrears built up, thus endangering the Society's reputation for speedy publication. Other difficulties (such as the fuel crisis early in 1947) prevented the arrears being overtaken in 1947, when only four, instead of six, numbers of *Monthly Notices* were published; by the end of 1947 the printing schedule was nearly a year in arrears.

In the Council's report for that year it was proposed that one of its nominees for the new Council be designated Editor of *Monthly Notices* and it was explained thus:

> The proposal to have a separate Editor of *Monthly Notices* is, on the other hand, a new one for recent times, though there is good precedent for it in the earlier history of the Society, no less a person than Professor Arthur Cayley having held such an appointment from 1859 to 1881. Since 1881 the duty has been reckoned amongst those attached to the office of Secretary and has in fact been rightly regarded as the foremost of those duties. However, under present conditions the Secretaries' other duties have become so heavy that the Council deems it essential to give them some relief. But the Secretaries will continue to deal with communications received until the Council reaches its final decision regarding their publication. The new proposal is to delegate to another member of the Council work subsequent to that stage, and not, of course, to delegate the responsibilities laid upon the Secretaries by the existing Bye-laws.

Flora M. McBain (later Secretary) who, as a member of the staff of HM Nautical Almanac Office had much experience of editorial work and of printing, was so nominated and duly elected in 1948 February. In the following year the Council was able to report that, 'It was with considerable satisfaction that the Council was able, in the middle of the year, to announce that all arrears of printing of *Monthly Notices* and *Geophysical Supplement* had been dealt with. . . .'. That announcement represented an enormous effort not only by the Secretaries and Editors but also by the new printers; it was, by any standards, a remarkable achievement.

In spite of the increasing amount of material it was decided to continue with six numbers a year right through to Volume 109 (1949).

The Editor of the *Geophysical Supplement* (to *Monthly Notices*) who was the 'Geophysical Secretary', had far less material with which to deal; from the end of the war only five numbers (Nos 5–9 of Volume 9) were published.

Occasional Notes was temporarily suspended after No. 9 (issued in 1941), but in 1944 the Council accepted a suggestion (from Spencer Jones) that it be revived in order to make available to Fellows translations of some of the early books on astronomy belonging to the Society. The first such translation, by Professor J.F. Dobson assisted by Professor S. Brodetsky, (which was available in manuscript) was to be of the

Preface and Book 1 of *De Revolutionibus Orbium Coelestium* by Copernicus which, surprisingly, had not previously been published. However, owing to editorial and printing delays, it was not published until 1947 May, as No.10. No.11 (*J.C. Adams and the Discovery of Neptune*, to which reference has already been made) quickly followed in August. In November the Council proposed that *Occasional Notes* should, from 1948, be issued at least three times a year; and in 1948 February it appointed an Editor (F. Hoyle — after Atkinson had at first accepted and then declined) and an Editorial Board to implement that policy. No.12 (by Hoyle, himself, with R.A. Lyttleton) was published in 1948 September and No.13 in 1949 September, but it was to be three years before No.14 was published. The Editor reported to the Council that he had much difficulty in obtaining suitable articles; they were either too technical (one was, in fact, published as a Council Note on Astronomy) or they overlapped with or duplicated material published elsewhere, especially in books and commercial journals.

Originally *Occasional Notes* was intended for Fellows only and was not on sale to the public. A special exception was made for the magnificent article on *Edmund Halley as Physical Geographer, and the Story of his Charts* (by S. Chapman, in No.9), for which a separate edition was published. In 1944 the Council decided that numbers containing translations of early books should be put on sale, and this was done for No.10 — but not for No.11. From No.12 onwards all numbers were 'published' (instead of 'issued') and put on sale, although the Council decision to this effect was not made until 1949 November.

6. Medals and Awards

It had become the custom to invite the Gold Medallist, if not a British subject, to give the George Darwin Lecture, and this custom was followed in 1939 when Bernard Lyot gave a memorable address on his coronagraph. However in 1940 the Gold Medallist, Edwin P. Hubble, had then to decline the invitation; it was to be another 12 years before the Society could hear, at first hand, the story of the fundamental discovery that could be said to have opened up the Universe. No Gold Medal was awarded and no George Darwin Lecture given in 1941 or 1942; but, as mentioned earlier, the Gold Medal was awarded to Harold Spencer Jones in 1943, the special procedure when nominating a member of the Council duly being observed. Svein Rosseland, the distinguished Norwegian astronomer who was in the United Kingdom, gave the George Darwin Lecture. By 1944 the Council decided that it could resume normal procedure and awarded the Gold Medal to Otto Struve; but he was unable to accept the invitation to give the George Darwin Lecture. Joseph Proudman, geophysicist and authority on tides, was accordingly invited and gave an absorbing address on tides in the Atlantic Ocean; he was the first, and appropriately so, resident British subject to give the George Darwin Lecture. For the years 1945 (Bengt Edlén, 'The Origin of the Emission Lines in the Solar Spectrum'), 1946 (J.H. Oort, 'Some phenomena connected with interstellar matter'), 1947 (M.G.J. Minnaert, 'The Fraunhofer lines of the Solar Spectrum') and 1948 (Bertil Lindblad, 'The dynamics of the stellar systems') the old custom was followed and the Gold Medallist also gave the George Darwin Lecture. In 1949 when the medal was awarded to Sydney Chapman, and the usual restriction was applied, it was possible to re-invite Otto Struve to give the lecture.

When the medallist did not also give the lecture it was the custom for the Presidential Address to be devoted to a comprehensive review of the medallist's work; this was done in 1941 (Plummer on Hubble) and in 1950 (Smart on Stebbins, though later in 1950 Joel Stebbins gave the George Darwin Lecture). These addresses and the texts of the George Darwin Lectures, together with the shorter addresses on the presentation of the awards, are published in *Monthly Notices*.

The Jackson Gwilt Medal and Gift was awarded three times during the decade: to R.L. Waterfield in 1942; to H.W. Newton in 1946; and to A.M. Newbegin in 1949.

Although no Associates were elected until 1945, a total of 13 were so honoured during the decade.

7. Library

Until 1945 only the most essential attention could be given to the library, although it continued to provide an almost complete service and was much used. No binding could be done and there were obvious gaps in the runs of foreign periodicals. However, due to the generosity of a Fellow, W.H. Owston, the Society was able to purchase, as they became available, a number of early books on astronomy; also several gifts and legacies in the form of books were received. Due mainly to shortage of staff (there were no staff specially allocated to the library until 1948) very little more could be done in 1946 and 1947. Such time as could be made available from the day-to-day routine was rapidly consumed in filling wartime gaps and reorganizing exchanges. Some binding was done but the binders were very slow, resulting in lengthy non-availability of periodicals. The Library Committee, under the chairmanship of H. Dingle, worked hard during these years, with little to show for its efforts. It reached decisions on the future organization of the library, as regards both classification and arrangement, and obtained the Council's approval to two comprehensive reports, one in 1948 October and the other in 1949 March.

The return of Miss Wadsworth, even if only on a part-time basis, to the library early in 1948 made it possible to make progress with these plans; she had been in charge of the library before 1939 and had a unique knowledge of its organization and content. The Council, in its *Annual Reports*, said (for 1948):

> Preliminary steps have been taken in the re-arrangement and re-cataloguing of the Library and a cabinet for the classified catalogue has been purchased. It has been decided to adopt the Library of Congress system of classification for the catalogue and the World List of Scientific Periodicals system for cataloguing periodicals in the Society's Library.

and for 1949:

> Much of the preliminary work of reorganization has now been finished, and at the end of 1949 the Library Committee was in a position to invite the cooperation of a panel of some two dozen experts to advise the Librarian on the preparation of the new classified catalogue. Their reports on the various subject sections of the library will include amended classifications and cross-references, recommendations to discard unsuitable works and duplicates, and proposals to make additions.
>
> The physical reorganization is a task for the future, but when it does take place it will inevitably add to the existing heavy demand on shelf space for normal accessions.

The 1949 *Report* goes on to give details of the many steps already taken to make better use of existing storage space and shelving.

Three further reports from the Library Committee, mainly on the physical problems of accommodation, had been considered at the Council meeting on 1949 October 14; but the most important content was the announcement of the offer of the Kuffner Library to the effect, 'Stephen Kuffner, son of the late owner of a private observatory in Vienna (the von Kuffner Observatory) wishes to present his father's library to the Society. . . . The Library, which is reported to be extensive and valuable, is stored in Vienna in about 70 boxes; the contents are not catalogued or known but there is nothing later than 1918.' The Council accepted the committee's proposal that the offer be accepted, subject to the acceptance of a suitable understanding regarding the disposal of duplicate items and satisfactory arrangements for transport. This promised to be the most valuable gift to the library during the decade; the many others, some of considerable value, are mentioned in the *Annual Reports* and recorded in the Society's archives.

Much work and effort was devoted, from 1945 onwards, to restoring exchange agreements with foreign societies and periodicals. The Council's policy, essentially that astronomers everywhere should have access to the Society's publications, is easy to state but difficult and time-consuming to implement. Many channels of communication had to be used, often through personal contact. But it was possible, for the first time for many years, to prepare an Exchange List in 1947; and thereafter not too difficult to maintain and extend it.

The usage of the library remained remarkably constant after 1945, averaging about 500 books borrowed each year by about 120 different borrowers, with rather less than 100 visitors a month for consultation.

8. The Society's Premises and Property

The Society's premises emerged from the war years essentially unscathed; but it was many years before the minor damage (mainly to windows) could be completely restored and the rooms could be redecorated. Apart from the changed usage of the former flat, partly for the British Astronomical Association and partly for library storage, the only major change was the conversion in 1946 of the outer office, previously used mainly as a cloakroom, to provide more office accommodation. Provision was made for the more secure shelving for the older, and more valuable, books in the library, and some of the benches (how uncomfortable they could be!) in the Meeting Room were repaired and recovered. The Society's valuable property, including stocks of publications, was returned from storage at Oxford and Cambridge. The pictures, including the two portraits (of the Reverend Lewis Evans and the Reverend William Pearson, one of the founders of the Society) presented to the Society during the war, and the series of Presidential portraits in the Meeting Room, were rehung; and the Harrison Regulator Clock again complemented the magnificent display of the early astronomical books, in their colourful bindings, in the Grove-Hills Library. Although dim, somewhat bedraggled, and in many ways inconvenient, the rooms continued to provide a dignified home for the Society. The boiler for the central heating system failed in 1947 November, resulting in some dislocation of meetings during the winter (the December meeting was held in the rooms of the Linnaean Society); a new boiler was installed during the summer of 1948.

The Society continued to provide facilities for the meetings of its guest societies, the British Astronomical Association and the London Mathematical Society, as well as for some committees of other organizations.

Most of the Society's instruments, ranging from large telescopes to small pieces of equipment, were on loan to institutions or Fellows at the outbreak of war and, with a few minor changes, the loans were renewed from year to year. After the death of Reverend T.E.R. Phillips some instruments were stored at Headley but in 1948 the Coleman 8-inch refractor was sold, making it possible to terminate the arrangements made in 1942 with the Rector and to dismantle the Observatory. A number of instruments were presented to the Society after the war and are duly recorded in the *Annual Reports* and in the archives.

The Photographic and Instruments Committee, mainly under the guidance of P.J. Melotte, had a difficult and frustrating task with instruments and, later, with restocking the Society's collection of slides and prints and adding films. For many years it proved impossible to find suitable firms able and willing to make slides and prepare prints up to the Society's standards. The Committee's strenuous efforts at last met with success so that, at the end of the decade, the wartime gap of six years had largely been made good and much progress had been made towards a stable procedure. In 1947 a small collection of slides (previously available for sale only) was made available for loan to Fellows.

9. The Society's Staff

As mentioned earlier, from the outbreak of war until the end of 1945, the whole of the administrative work of the Society was undertaken by the Assistant Secretary, with only occasional clerical assistance. Miss K. Williams, LRAM, after having served the Society with great devotion for 19½ years, retired due to ill-health on 1942 July 31; she was succeeded by the former Clerk, Miss E. Wadsworth MA, who had been released for other duties but was fortunately then available to return. From then until Miss Bower's appointment she had only part-time clerical assistance from Mrs U.M. Dewey. Eventually permission was obtained from the appropriate ministry to engage a full-time shorthand typist and Miss Jean Bower commenced duty on 1945 October 29 — and, owing to Miss Wadsworth's indisposition, performed with credit the Assistant Secretary's duties at the November meeting of the Society!

In anticipation of Miss Wadsworth's retirement, to provide her with some measure of relief, and to enable some of the accumulation of arrears to be caught up, the Council appointed a full-time assistant, Miss M.F.M. Garratt BSc, in 1946 November with a view to her succeeding to the post of Assistant Secretary during 1947. Mrs Dewey, after a month's overlap, ceased to work regularly for the Society though she continued to give occasional assistance whenever called upon because of illness or other cause.

Miss Wadsworth retired from the post of Assistant Secretary on 1947 August 17, after five years of unstinted service in conditions of unprecedented difficulty, and the Council placed on record its appreciation of her devotion to the interests of the Society. Miss Garrett was appointed to succeed her from 1947 August 18. However, the Council decided to create a separate post of librarian in order to overtake wartime arrears and to facilitate reorganization; Miss Wadsworth accepted the offer of the post on a part-time basis as from 1948 January 1. She devoted herself to these tasks with the same enthusiasm, energy and competence that she had shown as Assistant Secretary.

After two years' valuable service Miss Bower left in 1947 September, to be succeeded by Miss U.M. Samuelson, who proved to be yet another great asset to the Society. She shared with Mrs Dewey the same Christian name of Ursula and, perhaps inevitably, they became known affectionately to Fellows as Ursa Major and Ursa Minor. Following the tradition established by Miss Williams and Miss Wadsworth they contributed greatly to the happy atmosphere in the Society, where every Fellow and visitor was made personally welcome, and they did a great deal of excellent work for the Society. (Miss Samuelson later married a Fellow, C.A. Padgham.)

During this time the porter, R.J. Steel, and Mrs Steel, once again resident in their basement flat, continued to give friendly and competent service to the Society — not least appreciated being the provision of tea before meetings and for committees.

At various times (in 1942, 1944 and 1947) the Council considered the specification of conditions and terms of service for the staff, and the provision of adequate pension schemes. The existing contributory pension payments, to which the Society added 10 per cent of salary, were invested through an insurance company, did not apply to all staff and were generally inadequate. Miss Williams' pension was supplemented by the Society from its separate pension fund. There were obvious difficulties in standardizing conditions and pension schemes for a small staff, and several attempts had to be made before a reasonably satisfactory solution was reached.

Sadler, as Secretary from 1939 February (and Acting Secretary for a few months before) until 1947 February, had a special and inside knowledge of the work of the Assistant Secretary during those years. Miss Williams had not been well for some time and this undoubtedly affected her both in her work and personally; she was not the same happy and friendly person who had been such an asset to the Society. It is a great tribute to her that she, in spite of this, rose to such heights of achievement in 1939. Largely through the encouragement and guidance of Reynolds, she accomplished a prodigious amount of work; it was easy to decide to evacuate the office and store all valuable property but, especially in the conditions of 1939, it was not so easy to arrange and to do. As time progressed the work, at Oxford, became much less onerous and she coped admirably, especially with the transfer of material required for the meetings in London, but she realized that her ill-health would not allow her to take on the increasing workload after the return of the office to London, and she resigned. It is a great pity that she could not have completed her term as Assistant Secretary in less strenuous circumstances.

Miss Wadsworth, who had served so long as assistant and deputy to Miss Williams, seized the opportunity of appointment as Assistant Secretary to show her remarkable qualities. It is difficult to imagine anyone who could have done as much, and as competently as Miss Wadsworth, and, moreover, could have done it with such enthusiasm and obvious pleasure. She knew everything about the Society, rarely needed to consult minutes or files and believed in doing things immediately; and she made very few mistakes. She wrote to the Secretary several times a week: the letters were typed on the left-hand side only, leaving space for answers or comments on the right-hand side. She made herself available at times to suit the Officers, particularly until late in the evening on the Saturdays after meeting days, so that as much as possible of the 'post agenda' from the Council meetings could be dealt with. It was a great experience to work with such an enthusiastic, efficient and delightful person. The Council, and the Society, have shown their appreciation of her unexampled service and I too demand the privilege of putting on record my own indebtedness to her and the high esteem in which I regarded her. (Miss Wadsworth died on 1985 March 25, aged 99 years.)

D. Scientific Activity

1. General

It is not the function of the Society to engage directly in scientific research and it accomplishes its objective, namely the encouragement and promotion of astronomy, mainly by the provision of facilities for the exchange of ideas. The number of papers communicated to the Society provides some, albeit a crude, indication of the level of astronomical activity. This number, which fell from about 70 in 1939 to an average of only 36 in the years 1942–44, rose to about 65 in 1946, fell again to about 35 in 1947, and then rose to about 70 in 1949. In retrospect, it is impressive that after six years of war, the recovery of astronomy took place so rapidly as to make good the wartime 'disturbance' in so few years.

There were four main fields of scientific progress directly arising from wartime research that have had enormous influence on astronomy: radar and radio; electronics in all its aspects; automatic, programmable, computers; atomic and nuclear physics. By the end of the decade neither the electronic revolution nor the digital computer were sufficiently far advanced to revolutionize, as they subsequently did, instrumentation, the reduction and analysis of observations, and theoretical calculations; similarly, it was some time before 'nucleogenesis' replaced 'energy generation' in stars. However, radar was immediately applied to the determination of meteor orbits from the reflection (by day as well as by night) from meteor trails in the atmosphere, leading to the discovery of new streams and a much-improved understanding of the origin of meteors. The first such paper in *Monthly Notices* was in Volume 107 (1947) by J.P.M. Prentice, A.C.B. Lovell and C.J. Banwell ('Radio Echo Observations of Meteors'). The requirements of radio astronomy rapidly outgrew the equipment that could be diverted from its former wartime use, and its development had to wait for the design, financing and construction of radio telescopes. There were two papers in Volume 107, one on 'Solar Radio Noise', by C.W. Allen, and a purely theoretical paper by R.v.d.R. Woolley on 'Galactic Noise', arising from the wartime observations of J.S. Hey. However 'stellar' radio astronomy was not represented until Volume 110 (1950) with the first papers on 'discrete sources' and 'radio stars'.

The Council clearly accepted the 'new' astronomy by inviting J.S. Hey to contribute a Council Note 'Radio in Astronomy' to its *Report* for the year 1947. Earlier P.M.S. Blackett (later Lord Blackett OM, PRS), who appreciated the potentialities of radio astronomy more quickly than anyone else, had written privately to urge upon the Society the importance of being ready to publish work in this new field. It was at the suggestion of Blackett that, late in 1949, the Council set up a Radio Astronomy Committee; this Committee which included non-Fellows as well as Fellows under the chairmanship of W.M.H. Greaves, played a significant part in promoting the subject and in encouraging the provision of suitable equipment for its development.

Greaves reported verbally to the Council at its meeting in 1950 February. Council expressed appreciation of the work being done and agreed to endorse a resolution which would result in the continuation of the work. He reported in March on a further meeting, that had been held on February 27 in Edinburgh — the minutes of which had been circulated. Council adopted a resolution in the following amended form:

The Council of the Royal Astronomical Society strongly endorses the proposals put forward by the Physical Laboratories of the University of Manchester for the erection in the United Kingdom of a steerable paraboloid aerial of 250 feet diameter. The Council considers that by the erection of this apparatus the prestige of science in Britain would be considerably enhanced. In giving its support the Council places on record its opinion that the investigations to be undertaken are of high scientific importance including, as they do, the systematic survey of both the isolated centres and the general background of galactic radio emission, the study of the radio spectrum of the galactic and solar radiations, the extension of the meteor programme to meteors fainter than the 6th magnitude, the further investigation of auroral phenomena and the measurement of reflected pulses from the Moon, planets, gegenschein and solar corpuscular streams. The Council is impressed by the consideration that the construction of the proposed paraboloid would permit the continuation in the United Kingdom of new methods of astronomical research, which have been so greatly developed by the skill of scientists in the United Kingdom, and which are independent of climatic conditions.

The minute does not make clear to whom the resolution was addressed!

Greaves reported verbally on a meeting of the committee, held that morning, at the May meeting of the Council and submitted a draft resolution. The Council adopted the resolution in the following terms:

> The Council of the Royal Astronomical Society endorses the opinion of the Radio Astronomy Committee that it is highly desirable that radio observations should be made during the total eclipse of February 1952. The Council also endorses the opinion of the Committee that a basic programme of intensity observations at centimetric wavelengths be undertaken at this eclipse in order to establish whether limb-brightening occurs.

In the Report of the Joint Permanent Eclipse Committee (JPEC) for the year 1950 it was agreed to endorse, and finance, the Army Operational Research Group under J.S. Hey to observe radio noise at 1 cm and 10 cm wavelengths from a site within the belt of totality to determine the amount, if any, of limb-brightening. A more extended programme was earlier turned down by the JPEC as its estimated cost was £5000 — a sum that was (then) equal to the total annual income of the Society!

At its meeting on May 12 the committee heard Martin Ryle (later Sir Martin Ryle) speak on the work being carried out on radio astronomy at the Cavendish Laboratory, and it was unanimously agreed that the support of the committee should be given. An appropriate resolution was submitted to the October meeting of the Council, which was then adopted in the following terms:

> The Council of the Royal Astronomical Society has considered the work in Radio Astronomy which is being carried out in the Cavendish Laboratory at Cambridge. The Council understands that this work has now reached a stage where the research work must proceed side-by-side with routine observational work. It considers that it is in the interests of British Astronomy that both aspects of the work should be continued and extended during the next University Quinquennium, 1952–57, and strongly urges the provision of funds for that purpose.

It is not possible, in a short survey, to refer to the many outstanding papers that appeared in *Monthly Notices*: three only, each marking the beginning of some new line of thought, might be mentioned: H. Alfvén, 'Magneto-hydrodynamic Waves and Sunspots' in volume 105 (1945); F. Hoyle, 'The Synthesis of the Elements from Hydrogen' in volume 106 (1946). The third, which was perhaps to lead to more discussion than any other, was received by the Society on 1948 July 14, read at the Edinburgh meeting on October 29, and published in volume 108: namely, 'The Steady State Theory of the Expanding Universe' by H. Bondi and T. Gold (who was not then a Fellow). It is rare indeed that a single paper gives rise to so much interest, controversy, stimulation and fruitful research.

The revival of geophysical activity, as judged by the number and content of the geophysical papers received, was slower to start; it did not gain momentum until the next decade. However, the Society did take the initiative in one direction — perhaps too far in advance of the time! In 1947 October the Geophysical Secretary, B.S. Browne, reported that the Geophysical Committee recommended that a systematic survey be made of inland water resources and proposed that the Council should approach other organizations with a view to the formation of a joint committee on 'River Flow Surveys and Records'; however, there was little response and, in 1948 November, the project was dropped.

2. Eclipses

The one field in which the Society made a direct contribution was, with the Royal Society, through the Joint Permanent Eclipse Committee (JPEC) (F.J.M. Stratton was secretary for all, or most, of the decade). The Committee was responsible for co-ordinating eclipse expeditions from the United Kingdom, obtaining and allocating a governmental subvention, providing instruments (such as coelostats) on loan and (later) arranging logistic support from the Armed Services. It was impossible to send expeditions during the first part of the decade, but the eclipse of 1940 October 1 (which many astronomers had planned to observe) was observed by HM Astronomer at the Cape (J. Jackson) and his staff (see *Occasional Notes* No.9). The first total eclipse for which an expedition could be organized was that of 1947 May 20 in Brazil, for which J.A. Carroll planned to use an echelon spectrograph which had been shown to give excellent results in Omsk in 1936, but the expedition had to be abandoned after the aircraft carrying some of the observers, and the optical parts of the instruments, crashed at Dakar with the deaths of Alan Baxter and J.H. Strong (neither of whom was a Fellow); A. Hunter, a member of the Council and later to become Secretary, was badly injured.

Encouraged by the Secretary of the Society, in another capacity, the staff of the Specialist Air Navigation Course had planned a 'navigational exercise' to extend the duration of the total eclipse of 1945 July 9 (over Greenland) by flying along the track, but at the last moment the Air Council ruled that, although the war in Europe was over, the operation was not opportune.

3. The Council and Fred Hoyle

No historical survey of astronomical activity within the Society could be complete without reference to a 'phenomenon' that first came to notice in 1939 — namely Fred Hoyle. Although not then a Fellow (he was elected, on the proposal of Eddington, in 1940 May) he and R.A. Lyttleton (himself a prolific contributor) submitted a number of joint papers to the Society in 1939 and 1940; Hoyle later submitted many papers, either alone or jointly with others — there were about 25 in the years 1945–49. The early papers were primarily concerned with the origin of the solar system, with stellar structure and with the accretion theory of stellar evolution. In 1939 April Lyttleton had submitted a paper 'On the Evolution of the Stars' which, after adverse criticisms by referees, he subsequently withdrew; in partial replacement Hoyle and Lyttleton submitted, in 1939 October, a joint paper 'On the Accretion of Interstellar Matter by Stars.' It was this paper that led to an unfortunate lack of confidence between the authors and the Council that was to endure for several years. Since this had considerable influence on the functioning of the Council's refereeing system and since it affected the attitudes of the authors and their associates for a long time, some account has to be included here.

At that time almost all papers were discussed in the Council, and indeed, this usually occupied most of the meeting; referees' reports and authors' comments were read or summarized, and no paper was rejected unless two referees had so recommended. The Society took pride in the standard of the papers in *Monthly Notices*, and the Council, like any other editorial board, expected authors to co-operate by accepting those suggestions by referees that the Council itself considered to be desirable. In the case in point, however, the two authors were unwilling to make the revisions upon which acceptance was made conditional. A lengthy correspondence followed. Although many members of the Council had considerable sympathy with the authors, collectively they decided that the case was not one in which the Council should override the clear advice of the referees. Ultimately the authors withdrew the paper and published its substance elsewhere (*Proceedings of the Cambridge Philosophical Society*, volume 35).

The course of events is sometimes determined by incidental circumstances that are not appreciated at the time. The Council, although fully informed and responsible, has to deal with authors through some individual officer* and, in this instance, through the newly appointed Acting Secretary, Atkinson (he was not elected Secretary until 1940 February), who happened himself to have an interest in the subject of accretion. Atkinson later severely criticized the published paper in his

---

*Usually a Secretary. Responsibility for handling all stages of a paper was shared by the Secretaries, often on an agreed arbitrary basis — such as even-numbered papers to one Secretary with the odd-numbered papers to the other — but all papers would, if practicable, be briefly discussed by both Secretaries before the meeting of Council, and often with the President.

paper 'Accretion and Stellar Energy' in *Monthly Notices* Volume 100 (1940); this paper was, in turn, criticized by Hoyle and Lyttleton in Volume 101. Although the Council included some of the most distinguished astronomers of the day and had proceeded with the utmost impartiality, this circumstance appeared to leave additional doubts in the authors' minds.

There was a relatively small output of papers from Hoyle during the years when he was acquiring an enviable reputation in Admiralty research, for (among other accomplishments) the ability to diagnose and correct faults in electronic equipment. This was, however, a prelude to a flux of papers covering, in rapid succession, most aspects of astronomical research; a member of the Council remarked that, if this flow continued, there would soon be no field for other astronomers to explore. So many papers put a heavy load on the Secretaries, on the small number of referees that were available and on the Council itself. These early papers were characterized by a certain amount of speculative theory necessarily based on 'plausible' assumptions, and it was mainly these aspects that the Council, on the advice of its referees, queried. There was also some criticism of the structure of some of the papers which quoted and used the conclusions of earlier papers before these had been published and subjected to general discussion; and, although the Council gave authors much latitude as regards presentation, this did often fall short of the standards of *Monthly Notices*. Especially during paper-rationing much editing was found necessary; however, sometimes after lengthy arguments, almost all papers were accepted and published.

The situation was much improved when McCrea became Secretary in 1946; he spent a great deal of time in personal discussions with Hoyle who, a year later, was elected to the Council. Even so the impression lingered that the Council had been and was, unreceptive to the new ideas and approaches of the younger astronomers — a view which is not supported by the despatch with which the Society published the work on steady-state cosmology or, in fact, by a *post facto* assessment of accretion! Hoyle and his collaborators led the post-war generation to take a wide-ranging view of their subject, as did, for example, Jeans, Eddington, Milne and Chandrasekhar before the war; it has been the privilege of the Society to publish much of the pioneering work of all these authors. The minor frictions between the views of authors and referees, however irritating at the time, are trivial when regarded in the broad sweep of history.

4. The Royal Greenwich Observatory

One of the most significant astronomical events of the decade was undoubtedly the decision, and the beginning of its implementation, to transfer the Royal Observatory at Greenwich to Herstmonceux Castle in Sussex. Although the Admiralty would certainly have consulted the Society about such a step in the early years of its history and probably as late as the 1920s, no such consultation took place — and, frankly, the Society could hardly have expected to be consulted. The immediate effect on the Society was the increase, from 10 to 60 miles, in its distance from the largest cluster of astronomers in the country, coupled unfortunately with a serious break in the Observatory's observational programmes. However the long-term effects on British astronomy, and thus on the Society, have been far-reaching — as subsequent Chapters will show. In particular in the following quarter of a century there was a substantial increase in astronomical research activity at the Royal Greenwich Observatory, to give it its new name, and its staff members continued to play a very substantial role in the activities of the Society, supplying many Officers, Council and committee members, and Editors of the Society's journals.

CHAPTER 4

THE DECADE 1950–60

C.W. ALLEN

1. The World Astronomical Scene

This decade found astronomers caught up in events that were to change the horizons of mankind. The Society had to meet the challenge of rapid development and it should be judged by its success in doing so.

By 1950 most countries had just emerged from the difficulties caused by a devastating war. Some normality had been achieved and the war's aftermath was no longer seriously affecting the Society's activities. On the other hand some technical developments stimulated by the war had set up two new branches of astronomy that were increasing the astronomer's observing power.

(a) Sensitivity improvements in high frequency radio receivers had given rise to the radio telescope which was soon found to be capable of observing the Sun, the diffuse Galaxy, radio sources of surprising variety and a spectral line. The new subject of radio astronomy was well launched by 1950.

(b) Rocketry for astronomical observations had been developed to the extent that solar spectra had been obtained from above the ozone layer. Photography, spectroscopy and photoelectric recording, unimpeded by the Earth's atmosphere, could be anticipated at the beginning of the decade and were further developed in its duration.

No less important than the technical development was the panorama of the International Geophysical Year (IGY), 1957 July–1958 December. This covered several years of preparation, one and a half years of climax, and a lingering decline. It showed how scientific institutions and governments could, if pushed, carry out a worldwide co-operative programme. Astronomers might argue that the IGY name implied that it was to do with geophysics and that astronomers need not be involved, however it was soon found that the organizers regarded solar activity as an essential part of the IGY programme and very much under their command. This did not upset the Society because during this decade it was making a takeover bid for geophysics, but it upset some astronomers.

It was widely hoped that the launching of artificial satellites would be a crowning achievement of the IGY co-operative scheme. The first sputnik was indeed launched on 1957 October 4, nicely timed within the

IGY period. However, artificial satellite competition proved too severe to be kept within the IGY rules and a separate organization COSPAR was created in 1959 to deal with the prolific scientific activity arising out of rocket-propelled vehicles. The significance to the Society, and to astronomers, of the artificial satellite advent was that new areas of astronomy were developed and that traditional attitudes were forced to change.

In addition to radio and rocketry an influence was made on astronomy and geophysics by nuclear advances and even nuclear weapon tests. The stellar evolution problem became intimately dependent on nuclear reactions and geophysicists found a new source of seismic disturbances, exploitable by their experiments.

It can be seen that astronomy was increasingly stimulated by world events. The rate of discovery, the publication of papers, the variety of subject matter, the number of astronomers and the amount of observing equipment, increased throughout the period and showed evidence of a special acceleration from about 1958. The upsurge at that time may be regarded as the main astronomical movement of the decade. The repercussions could be noticed in astronomical popularity and education as well as in research.

The political developments within the British Commonwealth could not be neglected by the Society, although it will be found they had little effect. By long tradition any area once described as part of the British Empire would automatically regard the RAS as its natural astronomical centre. In this decade the British Empire melted away and the British Commonwealth of Nations, though it took over, was also in decline. However for astronomy all areas defined above had kept to the RAS, had sent observatory reports to the *Annual Report of Council,* and had shown no general sign of disintegration. We could say that the RAS coordinated British astronomy and the term British included other friends such as the Irish.

Naturally the Society made every endeavour to keep in the forefront of all the astronomical advances that characterized the decade. This presented no difficulty in radio astronomy because this was developed with great vigour within the British area. On the other hand the British contributions to rockets and satellites were not yet developed. The Society's contribution to geophysics was the commencement of the *Geophysical Journal* in 1958. This new journal which replaced the *Geophysical Supplement to Monthly Notices* gave increasing attention to satellites and the IGY.

2. The Society and its Presidents

The Royal Astronomical Society in this decade was composed of

Fellows, Junior Members and Associates. Although it is a learned society it made no special endeavour to ensure that the Fellows were astronomically qualified. The requirement for Fellowship, from the Society's point of view, was that those admitted had a *genuine* interest in astronomy or in geophysics. Many were dedicated to astronomy but that did not mean that Fellowship was in any sense restricted to professional astronomers. A search through a Fellows' list in this decade showed about half the Fellows to be listed at a private address containing no reference to an astronomically or geophysically connected institution. These would probably be considered amateurs. The other half would range from fully professional astronomers, such as the Astronomers Royal, to those who were attached to university departments of physics, mathematics, geology and meteorology, or to national laboratories. Most of this latter half would have taken a professional attitude to astronomy or geophysics and would have been considered professionals. The Society rather prided itself on its wholehearted acceptance of amateurs. It made a real endeavour to ensure that they took their part in the affairs of the day and elected them to some 10 or 20 per cent of the officer and Council member positions.

During the decade increasing attention was paid to geophysics as an accepted activity of the Society. The proportion of geophysical Fellows was increasing. However, the attitudes of astronomers and geophysicists showed some differences. The Geophysical Discussions were attended almost exclusively by professionals, whereas astronomy meetings had a good admixture of interested amateurs. Astronomy always held the type of public appeal that would urge amateurs to keep in touch by joining the Society and coming to the meetings.

The numbers of Fellows, Junior Members, and Associates increased during the decade from a total of just over 1000 to just over 1500. The rate of increase was slow at first (2 per cent per annum) but went up to 5 per cent towards the end and reached a spectacular 10 per cent in 1959. No doubt this was related to the worldwide interest in astronomy at that time.

Less than 10 per cent of the Fellows attended the monthly meetings. Most of these would live near London and could therefore expect to give and receive more than their share of the Society's efforts and benefits. About one-third of the Fellows lived overseas and when in England would expect to make their RAS pilgrimage, i.e. attend a meeting. The regular Fellows would be glad to see them and often they would be asked to say a few words at the assembly.

The various Presidents maintained the traditional decorum of the Ordinary Meetings. The first 10 or 15 minutes of each meeting was occupied with formal business which followed the well-tried pattern: Minutes, Announcements, Presents received, Candidates for Suspension (i.e. to have the name list suspended in the Library for all to see), Candidates for Election, Admission of Fellows, and Communications received. The President might then call on a visitor or two to address the meeting before commencing the main presentation of papers.

The essential duty of the President was to take the chair at meetings of the Society and Council. He would thus set the tone for the Society's activities, make all serious announcements, present medals, give the admission handshake to new Fellows, thank the speakers at meetings and control the Council.

The Presidents were, of course, formally elected by the Fellows at the Annual General Meeting but their selection was enshrouded in some mystery. This one procedure was taken almost entirely out of the hands of Council and left to the retiring President in consultation with other retired Presidents alive at the time. It was normal for a President to remain in office for two years; longer was not permitted by the Charter.

A President's main impact on astronomy was through the Presidential Addresses he would give at the Anniversary Meetings. These addresses were of two fairly distinct types which we will have to distinguish by calling them subject-addresses and medal-addresses. The first type would be given on a subject of the President's choosing and probably related to his own work or interest. The second type was on the award of a medal and therefore related to the work of the medallist. Early in the decade the Presidents used the two Anniversary Meetings to give one full address of each type. Thus about half the Gold Medallists would be recognized by a full address and all other medallists recognized more briefly. Moreover, a President could give only one subject-address during his term of office. The establishment of the Eddington Medal during this decade meant dealing with extra medals. The consequence was that all medal-addresses were shortened and then Presidents felt free to give two subject-addresses if they so wished. They also expected some assistance with the preparation of addresses required for medallists working in fields other than their own.

Without doubt Presidents were chosen for their promise of being able to preside over the Society's affairs and deliver relevant addresses, but the appointment would also have been made as an acknowledgement of their services to the Society and their eminence in astronomy or geophysics. Efforts to see that the honour went to those worthy of it would have faced some restrictions. From the nature of the Society's activities it was essential that the President lived within Great Britain and preferably not too far from London. Of the six Presidents mentioned in this decade, one lived in Glasgow, three in Cambridge and two in London. Services to the Society most likely to be taken into consideration for election would be represented by the offices of Treasurer or Secretary. It is interesting to note that 12 consecutive Secretaries between 1912 and 1947 all became President. Prior service on Council would be expected also. On the matter of eminence, the main visible distinctions would have been a Gold Medal of the Society, an FRS, and a knighthood associated with astronomy. Nevertheless the Society was desirous of having an occasional amateur in the chair and no strict adherence to the above services or distinctions could have been expected. We will find that the Society had every right to be pleased with its Presidents in this decade as in others.

It should be possible to glean from the Presidential Addresses how the President might have influenced the Society or astronomy. The medal-addresses would reflect the world astronomical situation since the medals were awarded on a world basis. Past Presidents had put much scholarship and eloquence into the medal-address occasions, particularly when the medal award was made the subject of a full address. Good examples are recorded early in this decade. However, the later method of honouring all medallists with shorter, more uniform addresses has probably given more satisfaction to the medallists, made a more effective acknowledgement of their work and thus kept up the status of the medal.

We must look to the subject-addresses in order to have the Presidents properly revealed. There were six Presidents at least partly associated with this decade and eleven Anniversary Meetings with their associated addresses (1950–60 inclusive). The Presidents' previous service, eminence, chairmanship and lectureship were above reproach. Their humility was surprising in that hardly anyone gave an address on his own work. They spoke of their times, their observatory, developments by others and co-operation, integration, or disagreement with others. We find two reviews of fundamental astronomy from a somewhat historical attitude, one on historical instruments, one on an observatory, one on the impact of a new cosmology, one on general questions of Earth formation and one on probability in astronomy. We must conclude from this that Presidents took a very broad view of their responsibilities and tried to enlighten the Fellows on astronomy as it was rather than promote their own subject.

At the beginning of the decade the chair was occupied by W.M. Smart MA, BSc, Regius Professor of Astronomy in the University of Glasgow. He had been Secretary 1930–37 and had served 11 years on Council. His career was divided between the Royal Navy, the Cambridge Observatory and the University of Glasgow. His astronomy was the natural outcome of a background in mathematics and navigation. At the time he was President he could be considered the senior teacher of astronomy in the country; he had written, or was writing, textbooks on spherical astronomy, stellar dynamics and celestial mechanics. His first Presidential Address (1950) was on the award of the Gold Medal to Joel

Stebbins for his development of physical methods in astronomical photography and for the results obtained by these methods. He took great interest in the development of fundamental astronomy and chose this subject for his second Presidential Address (1951) entitled 'The First Half of the Twentieth Century'. He acknowledged it to be a partial review as he wished to ensure that patient, fundamental astronomy was treated properly. He thought that there was no fear of the controversial modern fields of astrophysics and cosmology being neglected by astronomers of the day, and that these could pass with brief review in his address.

The next President was Herbert Dingle ARCS, DSc, Professor of History and Philosophy of Science in the University of London. He had been Secretary 1929–33 and had served 18 years on Council. His early work had been in spectroscopy, astrophysics, relativity and cosmology while he was attached to the Imperial College, London. He always had a philosophical turn of mind and could turn a neat philosophical argument. It need be no surprise that at the time of his election to office he was developing history and philosophy of science as an academic subject but he would have been regarded as a professional atronomer by those who knew him. His Presidential Address on 'Science and Modern Cosmology' was as arresting as any that had been heard in the Society's rooms. A group of Cambridge mathematical astronomers (see p. 143) had presented some highly original, highly controversial points of view on cosmology; Dingle attacked these views with high satire and compelled attention. He renamed the new 'cosmological principle' and 'perfect cosmological principle' as the 'cosmological assumption' and 'cosmological presumption'. Here, as in other places, Dingle provided stiff opposition to any facile or unprepared change. Any committee which included Dingle had to be very certain its motives were right before it could proceed to smash the status quo.

John Jackson CBE, MA DSc, FRS, had been HM Astronomer at the Cape until he retired in 1950. He had been Secretary 1924–29, had served 12 years on Council, and since retirement had prepared the General Index (up to 1950) of the Society's publications. Not long before his appointment as President he had been awarded the Gold Medal for his work on stellar parallaxes and his contribution to the general problems of star positions and proper motions. His life had been spent as a fundamental observing astronomer and his appointments included First Assistant at Greenwich as well as HM Astronomer at the Cape Observatory. His address (1955) on 'The Need for older branches of Astronomy' dealt with the questions that surrounded his life work: double stars, parallaxes and meridian astronomy. The Society was made to appreciate how much of fundamental astronomy was dependent on persistent, careful and forward-looking observing.

Sir Harold Jeffreys DSc, FRS, appointed President in 1955, was the Plumian Professor of Astronomy in the University of Cambridge. He had served on the Council for 11 years. He had been awarded the Gold Medal (in 1937) for his researches into the physics of the Earth and other planets and for his contributions to the study of the origin and age of the Solar System. He was the most prolific contributor to the *Geophysical Supplement* over the whole period of its life, 1922−57, and then started invading the new *Geophysical Journal*. All this time he was putting his astronomical work into the *Monthly Notices*. It was President Jeffreys who set the pattern of reducing the medal-addresses so that he could give two subject-addresses of his own. These were entitled 'The Earth's Thermal History' (1956), and 'Probability Theory in Astronomy' (1957). Jeffreys' contributions to the physics of the Earth were so comprehensive that it was impossible to review any terrestrial subject without being confronted with his work. His lecture on probability was perhaps the only Presidential Address in this decade that was written about a President's own work. Jeffreys had written massive books such as *The Earth, Methods of Mathematical Physics* (with his wife), and *Theory of Probability*. He gave no sign of ever coming to the end of his original ideas or being unable to cope with them. Only when he wanted others to understand them did he meet any trouble.

W.H. Steavenson LMSSA, Gresham Professor in Astronomy, brought with him long experience of the skies, telescopes and of observing astronomy. He had served 16 years on Council and had also been Chairman on the Photographic and Instruments Committee. In spite of his title he would have been regarded by many as an amateur astronomer because he was in fact a medical doctor in private life (he did not actually have a doctorate). The Gresham College was an ancient, traditional establishment of the City of London. Its astronomy was concentrated into a week or two each year — somewhat like a summer school. Steavenson's astronomy was of the sort that was a great help to many Fellows of the Society. His address entitled 'The Instruments of Sir William Herschel' called on his fine understanding of telescopes and appreciation of Sir William Herschel. Most of his observing was visual and in later years concerned with the photometry of old novae. Plate 11 shows Dr Steavenson at the Northumberland telescope at Cambridge about two years before he became President. It seems quite probable that he will prove to have been the last amateur President of the Society.

The decade finished under the Presidency of R.O. Redman MA, PhD, FRS, Professor of Astrophysics and Director of the Observatories, Cambridge. He had served a short spell, 1948−49, as Secretary and had been on Council for 5 years. After an academic mathematical training he had devoted his life to perfection in astrophysical observing at the

Plate 11. Dr W.H. Steavenson (President 1957–59) at the Northumberland telescope Cambridge about 1955. (Steavenson Papers, © Royal Astronomical Society.)

[*facing page* 154]

telescope. These activities had taken him from Cambridge to Pretoria and back to Cambridge. He was one of the few large-telescope men in the country and, as such, was in great demand for organizing large-telescope projects. It was perhaps natural that he should use his first Presidential Address to speak on the 'Work of the Cambridge Observatory', the leading university observatory of the country.

These pages have shown that the Society and its Presidents have upheld the standards developed over 130 years of history. However, the Society was instituted for the encouragement and promotion of astronomy and we will need to look much further into its activities and the astronomical researches that have passed through its hands to assess what progress had been made. These depend on management, the influence of the Society in the scientific life of the country and on the personality and scientific output of groups and individual Fellows. We will take up such matters in the following sections.

3. Management of the Society

The main activities of the Society came to life at the regular meeting days on the second Friday of each month from October to May. Many Fellows devoted these days fully to matters connected with the Society. The Officers attended preparatory meetings in the morning, the Council in the afternoon, the Ordinary Meeting in the late afternoon, and, somewhat worn out, the RAS Club Dinner in the evening. The members of Council and other Fellows would use available parts of the day for secondary meetings, formal or informal.

The regular business of Council was carried out in accordance with a fairly strict schedule which, for most of the decade, ran as follows:

October	Eddington Medal — nominations
	Business of the recess
November	Gold Medal — nominations
	Eddington Medal — selection
	Proposals for new members of Council
December	Gold Medal — selection
	George Darwin Lecturer — nominations
	Officers and Council — nominations
	Reappointment of temporary staff
January	Gold Medal and Eddington Medal — confirmations
	George Darwin Lecturer — confirmation
	Draft Narrative Report to Council — approved
	Preliminary presentation of accounts
February	Final presentation of accounts
	Proofs of *Annual Report to Council*
	Applications for renewed loan of instruments

March	Associates — nominations
	Report on Annual Contribution defaulters
	Appointment of committees
April	Consideration of Staff
	Further report on defaulters
	Associates — selection and election
May	Associates — diplomas signed and sealed
	Consideration of content of *Annual Report*
	Arrangements for recess

These matters occupied about 40 per cent of the total time in Council. Some of the remaining time was used for incidental items but the bulk of it was spent on the acceptance or rejection of papers submitted for publication.

The main activities of the Society were summarized each year in a document known to the Council as The Narrative Report but printed under the title *Report of the Council to the Annual General Meeting.* These reports reflect the year by year trends and changes. They give the impression that much was attempted to meet various internal crises as well as the demands of advancing astronomy.

Some of the regular work has been in the hands of standing committees which report back to Council. The names of these committees speak for themselves: the Finance Committee, the House Committee, the Publications Committee, the Photographic and Instruments Committee, the Library Committee, the Geophysical Committee and the Radio Astronomy Committee. Many of the Society's developments have emanated from these.

The responsibility for attending to the essential business was in the hands of the *Secretaries.* Three of these were allowed by the Charter of whom two were elected to handle the main activities of the Council and Fellows while the third — the Foreign Secretary — made contacts with foreign astronomers in connection with awards, lectures and visits. There was also much to be done in connection with geophysical work and to this end a 'Geophysical Secretary' was appointed from 1952 who, however, did not have official Secretary status.*

What was expected of the Secretaries was that they should keep the Society's work turning over and out of trouble. A division of their work was made by giving the senior Secretary responsibility for the meetings of Fellows. They would back up one another on all preparatory and administrative work. When the senior Secretary resigned the other Secretary had usually sufficient experience to manage the Council. Actually Dr Alan Hunter was shuffled into the position of senior Secretary in 1949 almost as soon as he was appointed but he lost nothing in effectiveness or responsibility because of this. He was the

*This was a recurring problem as explained earlier (Ed.).

able administrator of the Society for seven years in spite of ill-health in 1954. He was aided by Miss Flora McBain of HM Nautical Almanac Office as second Secretary. She resigned in 1954 and in 1955 provided the romance of the decade by suddenly marrying Donald Sadler, earlier Secretary and later President. She was succeeded by Professor C.W. Allen who managed a rather short spell until 1957. Then in 1956 and 1957 vacancies for Secretary were respectively filled by Professor Hermann Bondi and Dr Michael Ovenden. These two formed a powerful combination which carried on for many years. They had to withstand the 1959 expansions of astronomy and geophysics as they erupted into the next decade.

The secretarial duties laid down by the bye-laws have not always been carried out personally by the Secretaries but may have been delegated to other Council members or to the Society's staff. It was quite rare in this decade for the Secretaries to read out the letters and papers presented to the Society — indeed if a Secretary did this, it was because he or she happened to be the most available Fellow for the purpose. The Secretaries did remain the formal editors of the Society's publications but, here too, much of the work was delegated to the staff or to appointed Council members. In 1950 we find the appointment of Mr Fred Hoyle as editor of *Occasional Notes* (until 1951), Dr Raymond Lyttleton as Editor of the *Geophysical Supplement* (until 1952), and Dr Gerald Whitrow as Editor of the *Monthly Notices* (until 1952).

In 1952 it was found that the geophysical activities of the Society required more than an Editor, they required a secretary. In consequence Dr Lyttleton's position was renamed the Geophysical Secretary and he occupied this office until 1960. His own work was more in astronomy than geophysics but he continued to handle geophysical papers and to arrange the Geophysical Discussions throughout the whole of the decade. He had a responsible part to play in the production of the new *Geophysical Journal* in 1958. However at that time two joint editors of the *Geophysical Journal* were appointed to deal with the much-increased editorial work. These were Dr A.H. Cook and Dr T.F. Gaskell, of whom only Dr Cook was elected to Council; Gaskell first joined Council in 1964 as the first holder of the new formal position of Geophysical Secretary. A new era of responsibility to geophysics was ushered in at this time and in due course some sort of parity was reached between astronomy and geophysics.

In order to maintain its standing in the world of astronomy the Society elected Foreign Secretaries who had a widely accepted international reputation and a record of good service. In this decade the choice fell very naturally on Professor F.J.M. Stratton and Sir Harold Spencer Jones. In both cases their office of Foreign Secretary formed fitting conclusions to their long service to the Society.

Finances were certainly as important in this decade as in any other and the Society leaned heavily on its Treasurers. At the beginning Sir Harold Spencer Jones was finishing a five-year run as Treasurer and consolidating a recovery from war circumstances. Professor L.M. Milne-Thomson took over for four years. He was a go-getter for his money's worth; he once made the Secretary return a piece of new equipment to a firm because they would not give a discount, then ask for it back again because it was wanted discount or not. Dr Gerald Merton saw the decade through (and much longer). He seemed to thoroughly enjoy handling the investments for the benefit of the Society and lived out the worst of the inflation troubles. He gave great support to the introduction of the *Geophysical Journal* in 1958.

During the whole period the Treasurers had to face steeply rising costs — particularly in printing. Nothing was neglected that might augment the funds. Fellows were encouraged to take covenants on their subscriptions, since in this way the saved income tax could accrue to the Society; they were discouraged from compounding in order that subscriptions be maintained; investments were turned to equities as a protection against inflation; grant-in-aid benefits were sought wherever appropriate; and, of course, any bequests were gladly accepted. However, these endeavours could not prevent the need for increases in the Society's subscriptions and the prices asked for the journals. Some improvement was achieved towards 1959.

The *premises* of the Society in Burlington House were held in esteem by the Fellows through long familiarity but were far from adequate for its activities. Hopes were high in 1950 that a new Science Centre was to be set up in London which would cater for the learned societies, but such hopes had to be abandoned a few years later. Thus it became necessary for the Society to improve the facilities in the existing building. The Isaac Newton Room, previously the lumber room, was prepared to house the newly acquired Palomar Sky Atlas. The furnishings and the Atlas itself were a gift from Mr Jack Miller. The seating of the main Meeting Room was in poor repair and was replaced by more attractive and comfortable seating in 1956. However, this made no difference to the number that could be seated, which was about 100. The attendances at the Ordinary Meetings throughout most of the decade were close to saturation point and it was evident that more would have attended had their entry been assured.

The Society's premises were shared, to some extent, by other societies. They were used for the regular meetings of the British Astronomical Association and also the London Mathematical Society. The BAA had the use of rooms for an office and library.

The *publication* of astronomical researches, reviews, data and information had long been regarded as one of the prime methods by which the Society should promote astronomy. At the beginning of the decade the publication of the *Monthly Notices of the RAS* was regular, the *Geophysical Supplement to the Monthly Notices* was less regular, the *Occasional Notes of the RAS* sporadic, and the *Memoirs of the RAS* even more sporadic.

The *Monthly Notices* issued six numbers a year, making a volume of about 700 pages, and so continued throughout most of the decade. There was a dramatic increase right at the end. The *Annual Report of the Council* occupied one of the numbers (No. 2, 3 or 4) each year. Russian abstracts to the papers were included at the end of each number during 1958–59 but then discontinued.

The *Geophysical Supplement* issued a number twice a year averaging about 65 pages per number, with nine and six numbers making up two volumes. There was a major change in 1958 with the launching of the new *Geophysical Journal* to replace the *Geophysical Supplement*. The new journal started with 400 pages per year in four numbers, i.e. a factor of 3 output increase. Russian abstracts appeared during 1958–59 but were discarded in 1960.

During the decade eight numbers of *Occasional Notes* were issued containing 21 articles on 300 pages. This publication closed at the end of volume 3 in 1959 and was replaced by the *Quarterly Journal* which was to contain all the non-research material to be published by the Society. This replacement does not appear to have been thoroughly understood at the beginning, for we read, '*Occasional Notes . . .* is now beginning to appear at a much more satisfactory rate' (*Q.Jl R. astr. Soc.* 1, 136, 1960), despite the fact that it had issued its last number.

The first *Memoir* since the war was issued in 1954 but for the rest of the decade papers of *Memoir* dimensions were fairly frequent. Six *Memoir* parts containing eight articles appeared in 1954–59.

In addition to its own publications the Society had a close tie with *The Observatory*. This magazine continued to report all the Ordinary Meetings of the Society and also many other matters and events of great interest to the Fellows. By a block subscription arrangement it was possible for *The Observatory* to be distributed to all Fellows.

This account does not give a true indication of the growth in publications. There was a gradual increase at all times and a startling increase towards the end of the decade. When due allowance is made for extraneous factors, it can be estimated that the astronomical research published by the Society nearly doubled between 1957 and 1961. In the same period the published geophysical research was trebled. These increases were matched by other astronomical and geophysical publications in Europe and America and represented a surge forward of the whole science. We can look to radio astronomy, rockets, artificial satellites, IGY and their influence for most of the explanation.

An increasing concern was felt about the time-lag between the presentation of papers and their final distribution in print. There were time-losing factors associated with refereeing, printing, proof correcting, movement of papers from one place to another and waiting in paper queues while the person concerned could get onto the job. From 1952 Dr G.J. Whitrow was no longer able to continue his editorial assistance. For the next few years the total delay time was normally about 12 months. To meet competition with other journals and give satisfaction to Fellows this needed reduction to six months, and it was seen that this would make great demands on referees, secretary-editors, the RAS office, as well as the printer. The matter was considered so urgent in 1955 that arrangements were made for the Society to employ an Editorial Superintendent who would concentrate on the handling of papers. When this arrangement had settled and other loopholes had been tightened up, the lag was reduced to the intended six months. However, the tighter schedule was easily upset and needed continued vigilance.

The *Photographic Print and Slide Collection* of the Society was maintained during the decade for loan and sale to Fellows. It was supervised by the Photographic and Instruments Committee, which became the Photographic Committee in 1957, as from that time the instruments needed no further supervision.

During much of this period the Society's prints and slides were made up in two lists. One catalogue issued in 1953 gave the Society's collection assembled from various sources over many years. The other catalogue was of the Mount Wilson and Palomar photographs obtained from the 48-inch Schmidt and 200-inch telescope of Mount Palomar. These were supplied by an agreement (*Monthly Notices* **111**, 136, 1951; **112**, 254, 1952; **113**, 279, 1953) whereby the Society acted as the European distributors.

The Meeting Room was adorned with a complete set of Presidents' portraits. There were other portrait collections not on display which are concerned with Associates, medallists and others. In an endeavour to keep these more complete, a request was made in 1950 for Associates and medallists to present their portraits. This idea does not seem to have found its way into the Standing Orders and this may explain why this collection was in some disarray.

A clearance of the Society's *instruments* was made in 1956. The instruments were little used, they were not modern, the facilities for handling them were poor and the storage space was wanted for other things. It was decided to hand nearly all of them to the British Astronomical Association which had an active Curator of Instruments and facilities for their distribution. The instruments concerned were mainly rather small telescopes, some of which were incomplete. It is clear that both the RAS and BAA benefited from the transfer.

Nearly all the instruments retained had some historic interest; many were on loan to Museums. The instruments belonging to the Society after 1956 include:

The Harrison Clock (Grove-Hills Library)
Two Beaufoy Clocks (Office and Staircase)
A microscope with large stage (Upper Library)
Celestial Globe (Grove-Hills Library)
Five W. Herschel mirrors
One complete telescope
One Hilger Solar Spectrograph

The steps taken in 1948–49 to rearrange and recatalogue the *library* were continued. Several experts had advised on the reorganization. The Library of Congress system was used for the catalogue and the World List of Scientific Periodicals system for the periodicals. This proved a longer undertaking than anticipated and the reorganization was not substantially complete until early 1954. During the whole period Miss Wadsworth had the reorganization in hand and for some years was assisted by Mr R. Kenedy; they both left in 1954.

In 1950 stocks of pre-1930 *Monthly Notices* were reduced to make more space available. When the surplus was offered to Fellows at package and postage cost the response was so high that the office staff could not cope and the surplus was put into the hands of book agents.

Early in the decade the Society subscribed to the Fair Copying Declaration. According to this the reproduction, for bona fide individual use, of single copies of papers published by the Society would not be treated as breach of copyright so long as certain safeguards were complied with.

The essential *staff* required for running everyday affairs included an Assistant Secretary (in charge of the office), a Librarian, and a resident Porter (usually with his wife). In the 1950s the ups and downs in staffing affairs revolved mainly around Miss E. Wadsworth, Miss M. Garratt and Mr E.C. Rubidge.

Miss Wadsworth had first joined the Society in 1923 as Clerk, had become Assistant Secretary in the war years and Librarian in 1948. She retired at the end of 1954 after reclassifying and rearranging the library but returned a year or two later to give further help. At the 1955 Anniversary Meeting she was presented with a memento in recognition of her long service; she was awarded the MBE early in 1956.

Miss M. Garratt had been appointed Assistant Secretary in 1947 and resigned in 1956. She was in charge of the office and of editorial work. Following a long custom she sat in at the Council meetings to record the proceedings for the minutes. She also sat in at the regular Fellows' meetings to record the formal proceedings, such as announcements, elections, papers read, etc. Editorial work was a demanding and time-consuming part of her duties and when the pressure was on in 1955 to achieve a faster flow of papers it was decided that new staffing arrangements were essential. An Editorial Superintendent was to be appointed whose primary responsibility would be to reorganize for editorial efficiency. When Miss Garratt heard of this plan she offered her resignation but was dissuaded from leaving at that time.

On 1956 January 2 Mr E.C. Rubidge was appointed Editorial Superintendent and commenced the editorial work in a separate office. However the situation was strained. Miss Garratt took legal advice and claimed that the Society had dishonoured her contract which stated she would be in charge of the office. From the beginning Mr Rubidge was asked to act as clerk to the Council, that is, he took over some of the Assistant Secretary's duties. Thus there was certainly some truth in the claim that Miss Garratt's position was varied. The essential thing from the Society's point of view was that the editorial work be accelerated and as far as this was concerned the manoeuvre was happily successful. However Miss Garratt finally resigned as from 1956 April 1 and Mr Rubidge became Assistant Secretary in her place.

The purpose of the monthly *meetings of the Royal Astronomical Society* was mainly for the presentation of original astronomical work or lectures but the first 10 or 15 minutes was always used for formal business as outlined earlier in the chapter. No-one could expect such a well-established custom to change within a decade and indeed no change was made. The habit of asking foreign fellows or guests to address the meeting continued but with a tendency to reduce the time available to them.

The nature of the scientific part of the meetings was made known to all fellows through the reports in *The Observatory*. It should be explained that from about 1950 there was an increasing tendency for the editors of this magazine to rely on the speakers to provide a script of what they were supposed to have said. The reported discussions reveal the on-the-spot ideas of the Fellows who had just heard the paper.

The February meeting combined the Anniversary Meeting with the Annual General Meeting but a change was to come in 1960 when the Annual General Meeting was delayed until March to give sufficient time between the end of the calendar year and the presentation of accounts.

In addition to the astronomical meetings there were a varying number (3 to 6) of *Geophysical Discussions* in each session. Usually one meeting would be set aside for the reading of papers while the others would be in the form of discussions on specific topics. The duration and form of these meetings was rather more variable than the astronomical meetings and they were sometimes held jointly with other bodies. Although the status of astronomy and geophysics tended to equalize towards the end of the decade the meeting procedures remained distinctive.

The precedent set in 1949 of having a meeting out of London in the summer recess was followed, with some breaks, through the 1950s. The typical out of London meeting would include an Ordinary Meeting, a Geophysical Discussion, a colloquium and a public lecture. The intention has been to provide a meeting of wider interest so that more Fellows could take an effective part.

The years, places of meeting, and inviting institutions were:

1950	Dublin	Royal Irish Academy
1951	Exeter	University College of the South West
1952	Leeds	University of Leeds: Leeds Astronomical Society
1953	Newcastle	University of Durham: Newcastle Astronomical Society
1955	Glasgow	University of Glasgow: Glasgow Astronomical Society
1956	Bristol	University of Bristol
1958	Keele	University College of North Staffordshire

It will be noticed that towards the end of the decade the meetings were becoming less regular. Sometimes summer meetings lapsed because of other events such as the Joint Eclipse Expedition to Sweden in 1954. However, the Glasgow meeting was held despite being soon after the Dublin International Astronomical Union (IAU) meeting of 1955.

On occasions the Society has organized other events such as the 1954 Eclipse Expedition to Sweden and the 1955 conversazione before the Dublin IAU.

CHAPTER 5

THE DECADE 1960–70

G.J. WHITROW

1. A Decade of Expansion

The considerable expansion in astronomical and geophysical research that occurred in the 1950s continued and became even greater in the following decade. Astronomy has long been a subject with a special appeal to many besides professional astronomers and the widespread reporting of the discovery of some remarkable new types of celestial objects in the 1960s led to a marked growth of interest in the subject at all levels from the popular to the severely academic, but particularly in the universities and other institutions of higher education in the United Kingdom where there was a great increase in the number of students and academic staff during the decade. At the same time it was becoming widely recognized that great advances were being made in geophysics. This growth of interest in both subjects was reflected in the membership and in the activities of the Society. There was a steady increase throughout the decade in the number of Fellows and Junior Members. There was a corresponding increase in the number attending meetings, with the result that by the end of the decade the Society's Meeting Room had become far too small to admit all those who wished to be present. Indeed, 1969 was the last year in which meetings of the Society were held in the Society's apartments in Burlington House. Since they had taken place there regularly for nearly a hundred years (the move from Somerset House having occurred in 1874), the close of the decade can be truly said to have been the end of an era in the history of the Society.

The increase in the Society's activities in this decade was heralded by the appearance of a new publication in 1960, the *Quarterly Journal*. It replaced *Occasional Notes*, which had been originated in 1938. For some time difficulties had been encountered in obtaining articles for the latter, partly because of its narrow circulation. The main reason, however, that led Council to introduce the new journal was its belief that the increasingly specialized character and growing number of papers in the *Monthly Notices* made that publication less and less suitable as a link between the Fellows of the Society and between the Council and the Fellows. It was therefore decided that in future *Monthly Notices* would be restricted to original research papers and that the *Quarterly*

Journal would contain not only articles of the type previously published in *Occasional Notes*, on specific astronomical and geophysical topics of general interest to Fellows, but also the brief reports of meetings previously given in *Monthly Notices*, the report of the Annual General Meeting with the Narrative Report of Council and the Treasurer's Annual Report, the Presidential Address, the addresses of the President on the award of the Society's medals, the George Darwin Lectures, obituaries of Fellows, and the Council Notes. Moreover, authors were to be encouraged to write for the new journal by the assurance of a wide circulation for their articles because it was to be distributed free of charge not only to all Fellows of the Society but to all subscribers to *Monthly Notices*. It was also decided to take the opportunity to fill a gap that some authors had been aware of in the range of articles published by the Society. They had tended to be either original papers on specialized topics or else impersonal and authoritative reviews. Authors were to be encouraged to express their views 'without being either wholly comprehensive or wholly original provided only that the article is intelligible and readable.' The Editors of the new journal were the Secretaries of the Society and an Associate Editor (to begin with this was D.W. Dewhirst, the last Editor of *Occasional Notes*). A good example of this new type of article which they wished to include was provided in the first issue, published in 1960 September, by F. Hoyle. He gave a connected account of his views on the origin of the solar nebula without attempting to be either comprehensive or impartial.

The growth of interest in all aspects of astronomy in the late 1950s led to an increasing inflow to the Society of research papers of a high standard and consequently to an increase in the amount of material accepted for publication. This caused Council in 1960 to end the custom of bringing out one volume of *Monthly Notices* a year but to close each volume when it had reached a suitable size. As a result, for the first time in the Society's history, two volumes appeared in one year and the time-lag between acceptance and publication of papers, which had become nearly a year, was halved. In 1963 it was decided to reorganize the publication of *Memoirs*, which had appeared very irregularly and had failed to attract much support from subscribers so that its circulation had been disappointingly small. Under the new system, *Memoirs* was regarded as a supplement to *Monthly Notices*, everyone receiving the latter being supplied automatically with each *Memoir* as it was produced. This meant that authors of *Memoirs* were assured of a wide circulation, the decision whether a paper should go into *Monthly Notices* or *Memoirs* becoming a purely editorial one, depending on which journal was considered the more suitable for the material concerned.

By 1964 the increase in the amount of material submitted for publication by the Society had grown to such an extent that two new administrative steps were taken. To reduce the delay in the publishing of a paper produced by refereeing it was decided to set up small Papers Committees, with Council and non-Council members, to reduce the number of referee's reports that needed to be submitted to Council for a final decision on publication of the papers concerned. Also, to reduce the burden on the Society's staff, a new post of Assistant to the Editors was instituted, the holder of this post having effective responsibility for seeing all accepted papers through the press, from the manuscript stage. The latter arrangement, however, soon proved to be unsatisfactory, and after a thorough analysis of the situation it was decided, in 1965, that the future of the Society's publications could be best assured by bringing them into association with a commercial publisher with international connections. An agreement was therefore made with Blackwell Scientific Publications Ltd, of Oxford, whereby they undertook to publish all the Society's journals, on the Society's behalf, on a commission basis. The responsibility for the contents of these publications remained with the Officers and Council of the Society. These arrangements came into force with volume 131 of *Monthly Notices*, volume 70 of *Memoirs*, volume 10 of the *Geophysical Journal* and volume 7 of the *Quarterly Journal*. As a result by the end of 1966 the backlog in the printing of papers was reduced, the average publication time for a paper in *Monthly Notices*, apart from papers which were delayed for revision by the author, becoming less than six months.

In 1967 some informal discussion took place at the General Assembly of the International Astronomical Union in Prague on the possibility of co-ordinating or combining the major European astronomical publications. A specific proposal to combine *Monthly Notices* with two other European journals was considered by Council and opinions were sought from former officers of the Society. Although Council felt that there was a definite need for more effective co-operation between astronomers and astronomical institutions in Europe, there was considerable opposition to any proposal that would lead to the cessation of publication of *Monthly Notices* or its submergence in any other journal. The proposal was therefore rejected. The following year Council welcomed the inauguration of the new journal *Astronomy and Astrophysics — a European Journal* and sent the congratulations and best wishes of the Society to the editors.

An interesting development in the editorial organization of the *Geophysical Journal* occurred in 1970. Dr D. Davies, one of its Editors, moved from Cambridge (England) to Massachusetts in January, and with the agreement of Council and the collaboration of Dr E.A. Flinn, another Editor in the USA, he established an American office of the journal for the receipt and initial handling of papers originating in North America. This had an immediate effect on the flow of acceptable material for the journal and reinforced its status as an international periodical of geophysics.

During the decade the total membership of the Society grew from 1467 to 2322, an increase of nearly 60 per cent. Pending the availability of the new Scientific Societies Lecture Theatre in Savile Row, it was arranged that, after the summer recess in 1969, meetings of the Society should take place in the lecture theatre of the British Academy in Burlington House. It was agreed that the old Meeting Room of the Society should be divided partly into additional offices and partly into a Fellows' Room available for casual meetings, discussions and reading. The growth of the Society and the expansion of its activities had led, however, not only to increasing pressure on the Society's space but also on the Society's time, there being a definite need to extend the time available for the presentation of scientific work at the Society's meetings. To cope with this it was decided in 1968 that the formal business of the Society which had traditionally been conducted at the beginning of each Ordinary Meeting should henceforth be presented in an abbreviated form. A much larger gain, however, was achieved by the occasional use of the period in the afternoon before tea and the Ordinary Meeting.

One activity in which the Society had long been involved ceased at the end of the decade. This was the Joint Permanent Eclipse Committee of the Royal Society and the Royal Astronomical Society, which was no longer needed because the advent of space astronomy had made eclipse expeditions less important.

The decade which comprised the discovery of the quasar and the pulsar and of plate tectonics closed with the first landing by man on a celestial body. As we look back on those momentous years in the history of astronomy and geophysics, it is gratifying to find that the Society made its own contribution to the expansion of these subjects, for, by extending its activities to cope with the increasing flood of information and research that characterized the period, it maintained its reputation as one of the major institutions concerned with the critical evaluation and dissemination of the latest knowledge in these branches of science.

2. Presidents of the Society

During this decade it remained the custom for each President of the Society to hold office for two years. Because Professor R.O. Redman, FRS was elected in 1959, there were six Presidents during the 1960s. They comprised two optical astronomers, one radio astronomer and three theoretical astronomers (of these one was the Superintendent of

HM Nautical Almanac Office, the other two being astrophysicists). Between them they spanned the whole range of the Society's interests with the notable exception of geophysics, as is clear from their Presidential Addresses.

Whereas Redman's first Presidential Address had been on the work of the Cambridge Observatories, of which he was the Director, his second Presidential Address delivered on 1961 February 10 was on the subject of 'Photometry in Astronomy'. In the course of this address, at the risk of being considered 'hopelessly old-fashioned', to use his own expression, he defended the systematic accumulation of careful measurements of the brighter stars in the sky as an essential contribution to the foundations of astronomy. He welcomed the trend towards automation but stressed that, since the primary aim of photometry is to increase our physical knowledge and understanding of the stars, more precise knowledge of the intensity distribution in typical stellar spectra was needed before embarking on any large-scale project of stellar photometry.

The President of the Society from 1961 to 1963 was W.H. McCrea FRS, Profesor of Mathematics in the University of London at Royal Holloway College, well known for his theoretical researches in many branches of astrophysics ranging from the Sun to the confines of the Universe and also as an expert in relativity. He had frequently served on Council and was one of the Secretaries of the Society from 1946 to 1949. He had also been a Vice-President several times. He gave two notable Presidential Addresses to the Society. In the first, delivered at the Anniversary Meeting in 1962, he spoke on 'Evidences of evolution in astronomy'. He presented a valuable, wide-ranging survey of the developments that had led to the current state of knowledge of stellar evolution. In particular, the study of the HR diagrams of clusters had led to the crucial ideas that all the stars of a cluster have the same age and that the turn-off point from the main sequence indicates that age. He stressed that this discovery was something entirely new in astronomy. Although generally analogous to the discovery of the stratification of fossils on the earth, it was more momentous and more unexpected. Nevertheless, many unsolved problems remained, particularly of mass-loss from stars and the reconciliation of ages from galactic research and extragalactic research. He looked forward, however, to our being able to obtain in due course a similar diagram for galaxies as for stellar clusters and to the exciting possibilities of applying the results to problems of cosmic evolution. In his 1963 Presidential Address McCrea gave a synoptic review of cosmology, in the course of which he discussed the possible influence on our ideas of the universe due to a relation between the deficiency of information about distant regions, because of the finite speed of light, and the red shift in the radiation from those regions.

In 1963 Sir Richard Woolley FRS, was elected President. Before his appointment as Astronomer Royal in 1955, in succession to Sir Harold Spencer Jones, he too had spent some years in the southern hemisphere, in his case at Mount Stromlo Observatory of the Australian National University. It was therefore appropriate that his Presidential Address in 1964 should be on 'Astronomical observations in the southern hemisphere'. He pointed out that the Magellanic Clouds provide the best place to study a large collection of stars as a whole, but they can be viewed only in the southern hemisphere. He drew attention to the negotiations that were then proceeding concerning the erection of larger southern telescopes but he also said that there was still much useful work that could be done in this country. In particular, the Isaac Newton Telescope could be of great benefit for the future of British astronomy. The following year he devoted his Presidential Address to the subject of 'The RR Lyrae variable stars' and gave an account of both their kinematical and physical properties.

The next President was T.G. Cowling FRS, Professor of Applied Mathematics in the University of Leeds, distinguished for his researches in many branches of astrophysics, including applications of electrodynamics and his pioneer work on the construction of stellar models employing known energy sources. In 1966 he devoted his Presidential Address to a critical review of 'The development of the theory of stellar structure'. In a penetrating discussion of the work of the pioneers in this field he remarked that 'Eddington had the gift, characteristic of genius, of being able to reach correct conclusions by arguments at first sight dubious.' He also made a careful evaluation of Milne's work on stellar structure, concluding that its two most valuable features were that it influenced Chandrasekhar's theory of white dwarfs and that it had some useful mathematical consequences, since Milne was the first to introduce homology-independent variables. In concluding his review of these and subsequent advances in the subject, Cowling made the important point that in the future anyone working in this field who had been trained only as a mathematician or as a physicist or as an astronomer was going to be at a disadvantage. A new type of scientist, with his own specialized training will be needed. He hoped that current developments in theoretical astronomy at more than one university would supply such people. In his 1967 Presidential Address he turned his attention to the subject of 'Interstellar and interplanetary plasmas' and surveyed the field from the solar wind to cosmic radiation from galaxies. In particular, he stressed that observations of interplanetary plasma can be used to check the plausibility of interstellar and solar theories. He concluded by referring to space research and its links with astronomy. These often seemed to be slight, but in the field under discussion they were stronger. The motto of the Society is 'Quicquid nitet notandum' — 'Whatever shines is to be observed.' Radio astronomy, infrared astronomy and X-ray astronomy already translated 'nitet' as 'radiates' rather than 'shines'. Cowling appealed for particle radiation to be added to the list.

D.H. Sadler OBE, Superintendent of HM Nautical Almanac Office, a Secretary of the Society from 1939 to 1947 and General Secretary of the International Astronomical Union from 1958 to 1964, was elected President of the Society in 1967. His Presidential Address in 1968 was on 'Astronomical measures of time'. He claimed that this was the first occasion on which a President of the Society had taken the subject of 'time' for his address and he had chosen it to explain some of the mystique that had arisen in connection with Ephemeris Time and also to clarify the respective roles in astronomy of Ephemeris Time and Atomic Clock Time. Ephemeris Time is the independent time variable of dynamical astronomy, but because Atomic Time is far more precise it is more suitable for measuring the irregularities in the rotation of the Earth. Also, it provides this information much more quickly. The nature of the relationship between these two time-scales is of great theoretical interest. Atomic Time can be identified with the proper time of relativity theory and Ephemeris Time with co-ordinate time. Consequently, there is a minute relativistic gravitational variation between the two time-scales, although this is at present too small to be detected in the determination of Ephemeris Time. In his Presidential Address in 1969 Sadler spoke on 'Astronomy and Navigation', a subject with which to a greater or lesser extent he had been concerned for most of his life. His review covered bird navigation, conventional astronomical navigation, the use of artificial satellites and orbital and space navigation. He concluded that 'the astronomer is still very much in the navigational business', and that astronomy is therefore still a useful science in the practical sense.

The last President to be elected in the decade was the Director of the Nuffield Radio Astronomy Laboratories, Jodrell Bank and Professor of Radio Astronomy in the University of Manchester, Sir Bernard Lovell OBE, FRS. His Presidential Address on 1970 February 13 formed part of the celebrations of the one hundred and fiftieth anniversary of the foundation of the Society. Instead of speaking on his own field of research, he decided to speak on the general theme of 'The prospects for British astronomy'. He spoke particularly about the increasing measure of State interest in and control of the development of astronomy, as a result of which far more money was becoming available for the subject than ever before, even when due allowance was made for the effects of inflation. He pointed out that the current favourable situation existed because of the sympathetic attitude of the Science Research Council, but it was essential that there should be unanimity within and outside the Society regarding the objectives sought by British astronomers and the procedures for their operation. The main problems of finance and organization that had to be faced concerned observational astronomy and its extension throughout the electromagnetic spectrum. He reminded the Society that the absorption of radio astronomy and geophysics into its affairs had been a major factor in the development of these disciplines in this country and that the Society itself had been reinforced and strengthened by these extensions of its interests. The new forms of planetary science which were so speedily developing after the acquisition of lunar samples and the new terrestrial and space techniques for investigating the sky in the ultraviolet, infrared, X-ray and γ-ray regions presented a sudden and unprecedented opening of opportunities. Lovell stressed that it was the duty of the Society to embrace the new men, the new techniques and the discoveries in the same way that it had always done in the past.

The reference to the Science Research Council (SRC) by Lovell makes it appropriate to mention at this point that the SRC had been created in 1965 and that one of its first actions had been to take over the Royal Greenwich Observatory from the Admiralty. Although this had led, unfortunately, to the ending of the Board of Visitors to the Royal Greenwich Observatory, the setting-up of the SRC provided the opportunity for astronomy and geophysics to have access to more money than was previously available.

3. The Society's Medals and Special Lectures

In 1960 the Gold Medal of the Society was awarded for his outstanding contributions to astrophysics to V.A. Ambartsumian, Director of the Burakan Observatory, Armenia, USSR. In his address on this award the President, Professor Redman, referred to the principal researches of Ambartsumian, including his early expression of the view, later generally accepted, that a nova outburst involves only the surface layers of a star, his important contributions to the theory of radiative transfer, and his researches on stellar associations (loosely knit groups of stars of similar type and age). He was responsible for much of the inspiration that had led to the recent growth of Russian astronomy. The Eddington Medal was awarded to R. d'E. Atkinson, for his pioneering work in 1931 on atomic synthesis and stellar energy, and the Jackson Gwilt Medals and Gifts to F. M. Bateson and A.F.A.L. Jones, for their work on variable stars in the southern hemisphere. The 1960 George Darwin Lecture was delivered at the meeting of the Society in May by Ambartsumian. He spoke on 'The evolution of stellar systems' and laid particular emphasis on the cosmogonic activity of the nuclei of galaxies.

Recognition of Atkinson's pioneering work in 1960 was followed in 1961 by the award of the Eddington Medal to H.A. Bethe for his definitive identification of energy-generating processes in stars.

The Gold Medal for 1961 was awarded to Herman Zanstra of the University of Amsterdam for his work on gaseous nebulae, beginning in 1926 when he saw that atomic photo-ionization is the primary process by which absorption of stellar energy produces a nebular emission line spectrum. The George Darwin Lecture was given by Zanstra on October 13 with the title 'The gaseous nebula as a quantum counter'.

In 1962 the Gold Medal was awarded to Bengt Strömgren of Copenhagen for his contributions to stellar and interstellar astrophysics. Much of his work could be regarded as the interstellar analogue of the work of the previous year's medallist on planetary nebulae. When he came to receive his medal on October 12 he delivered the George Darwin Lecture on 'Problems of internal constitution and kinematics of main sequence stars'. The Eddington medal for 1962 was awarded to André Lallemand, Director of the Physical Astronomy Laboratory of the Paris Observatory, for his outstanding work on photomultipliers and particularly for his successful development of the electron camera.

The Gold Medal for 1963 was awarded to H.H. Plaskett, who had retired from the Savilian Chair at Oxford in 1960, for his outstanding contribution to solar physics. His work centred mainly on absorption line problems, problems associated with the mechanical stability of the solar atmosphere, and the demonstration of the existence of meridional currents on the Sun. Plaskett's father was also a Gold Medallist of the Society (in 1930) and there have been only two other families with more than one Gold Medal of the Society to their name, the Struves with four and the Herschels with three. The 1963 Eddington Medal was awarded jointly to Allan Sandage, of Mount Wilson and Palomar Observatories, and Martin Schwarzschild, Professor of Astronomy in the University of Princeton, for their theoretical and observational investigations of stellar evolution. The Jackson Gwilt Medal and Gift was presented to G.E.D. Alcock for his systematic searches for and observations of meteors and comets, and for the discovery of two comets. The George Darwin Lecture was given by R. Tousey in October on 'The spectrum of the Sun in the extreme ultraviolet'.

In the course of 1963 Council wished to recognize the growing importance of geophysics by instituting a second Gold Medal, to be awarded for work of a geophysical character and thereby avoiding the difficulties of comparing the claims of geophysicists and astronomers upon a single medal, but although the Society agreed that in future two Gold Medals should be awarded annually no restriction was placed on either. It was also decided to establish an annual lecture to be called the Harold Jeffreys Lecture on some subject of geophysical interest to the Society. A fund of £1500 was provided for this out of the profits from the sales of *The Earth Today*, the special bound version of volume 4 of *Geophysical Journal* commemorating Sir Harold Jeffreys' seventieth birthday. At the Geophysical Discussion meeting of the Society on October 25 Sir Harold Jeffreys himself delivered the inaugural Harold Jeffreys Lecture on 'How soft is the Earth?'.

The Gold Medals for 1964 were awarded to Martin Ryle, for his outstanding contributions to the development of radio astronomy by astronomical discoveries and by the invention of techniques and instruments, and to Maurice Ewing for his contributions to marine geophysics and the propagation of elastic waves. The Eddington Medal was awarded jointly to H. Friedman and R. Tousey for their pioneering work, by means of rockets and satellites, in ultraviolet astronomy. The George Darwin Lecture was given on October 9 by P. Swings on 'The spectra of comets', and the Harold Jeffreys Lecture was given on October 23 by Professor Ewing on 'The sediments of the Argentine basin'.

In 1965 a Gold Medal was awarded to G.M. Clemence in recognition of his applications of celestial mechanics to motions in the Solar System and his fundamental contributions to the study of time and the system of astronomical constants. Dr Clemence had served many years in the United States Naval Observatory but at the time of the award he was Senior Research Associate at Yale University, working at the Celestial Mechanics Center with an earlier medallist, Dirk Brouwer. In October Clemence gave the George Darwin Lecture, his subject being 'Inertial frames of reference'.

A Gold Medal was also awarded to Sir Edward Bullard for his pioneering work in geophysics and, in particular, his leadership in the field of studies of the geology of the ocean floor. Among his many achievements was a theoretical study which showed that the main terrestrial magnetic field could be accounted for by dynamo action in a fluid conducting core. He and the 1964 medallist Ewing were the great pioneers of modern geophysical studies at sea, and he had recently been one of those producing definite evidence of continental drift. The Harold Jeffreys Lecture was given by Bullard in December on 'Electromagnetic Induction within the Earth'. The Eddington Medal was jointly awarded to R.V. Pound and F.A. Rebka Jr for their decisive experiment for measuring the gravitational red shift on the surface of the Earth through the methods of nuclear physics. They succeeded in detecting the shift produced by the change in potential over a height of only 74 feet and found that the shift given by Einstein's general theory of relativity was confirmed to within 5 per cent.

For his work on spectroscopy and optical astronomy, especially in relation to the construction and erection of the 200-inch Palomar Telescope, a Gold Medal was awarded to I.S. Bowen in 1966. As the President, Professor Cowling, pointed out in his address on this award, it was largely as a result of Bowen's personal contribution and leadership that this great telescope came into operation for regular scheduled observations early in 1949 November. The other Gold Medal for 1966 was awarded to the famous physical chemist and Nobel prize-winner H.C. Urey for his work on cosmo-chemistry, particularly as applied to the problem of the origin of the Solar System. He was the first to apply a thorough discussion of the chemical data to this problem, which led him to conclude that the planets began as cold, accreted objects. The Eddington Medal for 1966 was awarded to R. Wildt for his identification of the negative hydrogen ion as providing the chief source of opacity in cool stars such as the Sun. The George Darwin Lecture was given in October by Dr Bowen, who spoke on 'Future tools of the astronomer'. Professor Urey gave the Harold Jeffreys Lecture in September, his subject being 'Abundances of the elements, with special reference to iron'.

Hannes Alfvén, of the Royal Institute of Technology, Stockholm, was awarded a Gold Medal in 1967 for his fundamental work on cosmical electrodynamics, of great importance in regard to many problems in geophysics and astronomy. As Cowling remarked in his address on the award, the significance for geophysics and astronomy of Alfvén's development of magnetohydrodynamics as a separate subject can be gauged from the fact that one can hardly read a paper on interstellar, interplanetary or magnetospheric phenomena without finding a reference to Alfvén's waves and frozen-in fields. A Gold Medal for 1967 was also awarded to Allan Sandage for his fundamental work on stellar evolution and the history of the Galaxy and in providing optical data basic to modern cosmology. In addition to his work on stellar evolution from the main sequence to the red-giant stage, for which he had shared the 1963 Eddington Medal, Sandage had made fundamental contributions to the study of galaxies and cosmology, including an important correction to the Hubble 'age', and to the discovery of quasars. The 1967 Eddington Medal was awarded to R.F. Christy for his work on the non-linear theory of pulsating stars, which has enabled a close comparison to be made with observations of cepheids and RR Lyrae variables. The Harold Jeffreys Lecture was given on May 12 by Professor Alfvén, his subject being 'On the origin of the solar system', and the George Darwin Lecture was given at an additional Ordinary Meeting on September 8 by Professor Christy, who spoke on 'The theory of cepheid variability'.

In 1968 a Gold Medal was awarded to Fred Hoyle for his contributions to theoretical astrophysics, especially the theory of stellar structure, nucleosynthesis and cosmology. In particular, his theory of the stellar synthesis of carbon and heavier elements accounted for an impressive array of observational data that had not been explained in any other way. A Gold Medal was also awarded that year to W.H. Munk, of the Scripps Institute of Oceanography at La Jolla, California, for his distinguished contributions to oceanography and to the study of variations of the rotation of the Earth. As the President, D.H. Sadler, pointed out in his address on this award, it was particularly appropriate that Munk's work, which showed so clearly the direct relations between geophysics and fundamental astronomy, should be recognized by the Society. The Eddington Medal was awarded jointly to R. Hanbury Brown and to R.Q. Twiss, for their invention and theoretical study of the intensity interferometer, which has now led to accurate measurements of the angular diameters of a series of stars. The Jackson Gwilt Medal and Gift was awarded to J.G. Porter, in recognition of his work on the orbits of comets and meteors and of his activities in the promotion of public interest in astronomy. The Harold Jeffreys Lecture was delivered at the Geophysical Meeting on March 29 by Professor Munk, the title being 'Once again — tidal friction', and the George Darwin Lecture was given by Professor Hoyle at the Ordinary Meeting on October 11 on 'Highly condensed objects'.

A.T. Price, who had recently retired from the chair of applied mathematics in the University of Exeter, was awarded a Gold Medal in 1969, in recognition of his work in geomagnetism and especially for his studies of the electrical conductivity in the interior of the Earth. His work was of great significance for theories of the Earth's interior, particularly of the upper mantle. A Gold Medal for 1969 was also awarded to Martin Schwarzschild, Eddington Medallist in 1963, for his fundamental contributions to the theory of stellar structure and evolution, to astronomical observations from high altitude and for the lively inspiration that he had given to the astronomical community. Modern ideas on stellar populations and the evolution of galaxies have been profoundly influenced by Schwarzschild's work and he has also contributed to clarification of a wide variety of problems in astronomy and indirectly to many discoveries besides those associated with his name. The Eddington Medal that year was awarded to Anthony Hewish of the Mullard Radio Astronomy Observatory, Cambridge, for his discovery of the pulsating radio sources now known as the 'pulsars'. His work on the scintillation of radio sources through the interplanetary medium led him to construct a very large radio telescope, operating at the comparatively long wavelength of 3.7 m, which could be used to cover most of the sky each week. It was this telescope which first gave discernible signals from the pulsar CP1919, and subsequent discoveries have shown that no better instrument could have been devised for their detection.

As the President remarked in his address on this award, few astronomical discoveries can have excited astronomers, both professional and amateur, and indeed scientists in general as much as the revelation that our Galaxy contains this remarkable species of celestial clock, ticking away with such amazing regularity and at such short intervals. The 1969 George Darwin Lecture was delivered on October 10 by Professor Schwarzschild on 'Stellar evolution in globular clusters', and Professor Price gave the Harold Jeffreys Lecture at a Geophysical Meeting on October 24 on 'The electrical conductivity of the Earth'.

In 1970 only one Gold Medal was awarded, in this case to Dr Horace W. Babcock, the Director of the Mount Wilson and Palomar Observatories, for his outstanding contributions to astronomy in the investigation of the magnetic fields in the Sun and stars and also in the design and development of astronomical instruments. In fact two medallists were selected at the 1969 December Council meeting. The second was also an astronomer rather than a geophysicist and the award was not confirmed in 1970 January. This is the only time this has happened since the award of the medal to Einstein was not confirmed in 1920, but the first volume of the Society's *History* describes several such occurrences in the nineteenth century.

The Eddington Medal for 1970 was awarded to Professor C. Hayashi of Kyoto University for his work on the pre-main-sequence contraction of stars and in particular for demonstrating the existence of fully convective stars. The George Darwin Lecture was given on November 13 by Professor H.C. van de Hulst of Leiden Observatory on 'Cosmic ray electrons'. The Harold Jeffreys Lecture for 1970 was not delivered until 1971 March 12 by Dr Frank Press. His subject was 'The Earth and the Moon'. The lecture was of particular interest because it contained information based on the study of actual samples of material collected from the surface of the Moon.

4. Council and the Management of the Society

General responsibility for the smooth running of the Society and the efficient conduct of its affairs lies with Council. This consists of the President, four Vice-Presidents, the Treasurer, three Secretaries and twelve other members. During this decade an important change occurred in connection with the Secretaries. It arose from the increasing involvement of the Society in geophysics. For some years one of the members of council had been officially responsible for organizing the Geophysical Meetings of the Society but he did not have official Secretary status. By the early 1960s this was coming to be recognized as increasingly anomalous and steps were taken to rectify the situation. There was, however, a constitutional point at issue. According to the Society's Charter only three Secretaries are authorized. Two of these had been charged with responsibility for the main activities of the Society whereas the third, called the Foreign Secretary, had made contacts with astronomers in other countries in connection with awards, lectures and visits. On the recommendation of Council, it was agreed at the Annual General Meeting of the Society in 1963 that the Foreign Secretary should be replaced by a Geophysical Secretary as one of three Secretaries authorized by the Society's charter and that this should be effected by the modification required to Bye-law 38. In 1964 T.F. Gaskell thus became the first officially styled Geophysical Secretary of the Society. The title 'Foreign Secretary' was thereupon dropped and replaced by the title 'Foreign Correspondent'.

Another important change that occurred in this decade concerned the method of election to Council. This was under consideration by Council from 1963 when a special committee was set up to examine the question. Despite a divergence of view in this committee, it was eventually agreed by Council, in 1965 December, that when the extra-Council nominations for the 1966 Council were known, a Council list of nominations would be submitted containing fewer than the 12 names required for 'Ordinary Members of Council'. At its next meeting, on 1966 January 14, Council was informed that three extra-Council nominations had been received and after some discussion it was decided to submit to Fellows a Council list incomplete to the extent of one name under 'Ordinary Members of Council', thereby allowing Fellows to vote for an extra-Council nominee without having to vote against a candidate on the Council list. This practice was generally welcomed and was followed again in 1967 when Council decided to submit a Council list of only 19 names for the next Council, but six additional nominations were included. In establishing this new pattern of elections, it was decided to present the ballot list in alphabetical order with no separation between Council nominations and other candidates. The opportunity was taken to remind Fellows of their right to make proposals for Council members under Bye-law 8. Later in 1967 Council agreed that the next ballot paper circulated to the Fellowship be accompanied by an envelope for its return, addressed to the Scrutineers of the Ballot and with provision for the signature and the name, in block letters, of the voter. It was agreed that all nominations within the category 'Ordinary Members of Council' be merged in a single list arranged in alphabetical order. It was also agreed: that the report of the Scrutineers to the President on the result of the ballot should in future take a standard form, namely, an announcement of the total number of voting papers received, the number of such papers invalidated and the number of valid votes cast, followed by the names of the newly elected Council; and that the Scrutineers be asked to communicate confidentially to the officers and Council a report on any aspect of the ballot to which, in their judgement, the attention of officers and Council ought properly to be drawn, all on the understanding that the secrecy of the individual vote would be strictly preserved.

At the beginning of the 1960s the Society was fortunate to have as Secretaries Hermann Bondi, who had held office since 1956, and M.W. Ovenden, who had held office since 1957. The Foreign Secretary was R.O. Redman, who was elected at the Annual General Meeting in 1961 February, following the death of Sir Harold Spencer Jones who had been in office for the previous five years. In 1964 Bondi was succeeded as Secretary by F. Graham Smith, the first radio astronomer to be elected an officer of the Society. At the same time, as previously mentioned, T.F. Gaskell was elected Geophysical Secretary. As a result, C.W. Allen, who had succeeded Redman as Foreign Secretary in 1962, became Foreign Correspondent, a post he held until 1968, when he was succeeded by W.H. McCrea. In 1966 Ovenden was succeeded as Secretary by Derek McNally, who remained in office for the rest of the decade, as did Smith, Gaskell and McCrea.

Although according to the bye-laws, the Secretaries are the formal editors of the Society's publications, during the 1950s it had become necessary to delegate much of the work of editing to others, both on Council and on the staff. When the *Quarterly Journal* first appeared, in 1960, the Secretaries, H. Bondi and M.W. Ovenden, were named as Editors and, as previously mentioned, D.W. Dewhirst, of the Cambridge Observatories, was appointed Associate Editor, specifically charged with the duty of obtaining and editing articles of wide interest for it. He was succeeded in 1965 by C.A. Murray, of the Royal Greenwich Observatory. In 1968 Council decided to consider the question of rewording Bye-law 79 allowing only the appointment of the Secretaries as editors. It was pointed out that the widening scope and the mounting volume of the Society's publications had made this requirement impossible of fulfilment. Since there had been a number of instances in recent years where, with the approval of Council, the editorship of certain journals had been deputed to Fellows other than the Secretaries, the time had therefore come to take account of this situation in the bye-laws. Council therefore agreed to recommend at the next General Meeting of the Society that Bye-law 79 be modified so as to read:

> 79.   The Secretaries, together with such other Fellows as the Council may appoint, shall be the Editors of all the Society's publications, and, when council does not otherwise provide, they shall have the charge (under the direction of the Council) of printing and publishing the Memoirs, papers and other publications of the Society.

This was approved.

Another duty that had been laid on the Secretaries by the bye-laws was to read out at Ordinary Meetings of the Society the full details of each candidate formally proposed for Fellowship. Towards the end of the 1960s there was a growing need to increase the time available at the Society's meetings for the presentation of scientific work. As previously mentioned, one way of coping with this was to abbreviate the formal business conducted at the beginning of each meeting. In 1968 December Council agreed that, although Bye-law 14 made it obligatory for one of the Secretaries to read out full details of each candidate formally proposed for Fellowship, Bye-law 16 imposed no such requirement in the case of candidates recommended by the Council for election. Consequently, Council decided that full details of each candidate for Fellowship should continue to be read at Ordinary Meetings but that only the names of candidates recommended by the Council for election should be read in future. According to Bye-law 77, at Ordinary Meetings of the Society the Secretaries were also required 'to read the letters and papers presented to the Society.' Although Council agreed that the current interpretation of that requirement, represented by the reading of a list of authors and titles of such papers, was reasonable and should be continued, it was realized that in due course it might be necessary to consider the possible revision of Bye-laws 14 and 77 to provide a measure of flexibility.

The continuing expansion of research activity in astronomy and geophysics in the 1960s caused a large fraction of time at Council meetings to be devoted to the consideration of reports by referees on papers submitted for publication by the Society. At the beginning of 1964 Council set up a committee to consider the handling of these papers and make recommendations to Council with a view to improving the current arrangements and to devising an effective system that would relieve Council of the necessity for routine consideration of every paper received during the session, while allowing time for both Council as a whole and interested individual members of Council to consider more thoroughly such difficult cases as might arise. It was Council's belief that, although a refereeing system is essential for maintaining the high standard of the Society's publications by eliminating unsuitable papers and improving the standard of presentation of many accepted papers, it was important that the delay produced by refereeing should be reduced to a minimum. The committee reported back in March and Council accepted its recommendation that small Papers Committees, with Council and non-Council members, be set up to give more detailed and speedier consideration to reports on papers than was possible under the previous system. Two astronomical papers committees and one geophysical papers committee were constituted. The new system worked well for the trial period of one year, at the end of which it was agreed to merge the two astronomical committees into one committee.

The steady growth in the Society's membership and activities led to increasing pressure on space, with the result that for most of the decade the problem of finding additional accommodation was of major concern to officers and Council. Shortage of space in the Society's premises was most obvious to Fellows on the days of Ordinary Meetings but it also affected the office. By 1964 it became necessary, with the consent of the Library Committee, to use the Newton Room for the Assistant to the Editor, who required a quiet place in which to work away from the necessarily noisy General Office. In the following years it became increasingly uncomfortable to squeeze into the Meeting Room all the Fellows and guests who wished to attend meetings of the Society. Other Societies were similarly affected. Although some alleviation of the situation was hoped for after the forthcoming move of the Royal Society to Carlton House Terrace, a major requirement was for a large communal meeting room. Negotiations with the Ministry of Public Building and Works led to the preparation of a new lecture theatre in the Civil Service Commission Headquarters building at 23 Savile Row to be used by different scientific societies. The Royal Astronomical Society decided to contribute £1000 towards the cost of this. Meanwhile in 1968, as a temporary expedient, Council agreed to the experimental use of closed-circuit television, so that an overflow meeting in the library could follow the proceedings taking place in the main lecture theatre below. As a consequence of the move of the Royal Society, a reallocation of accommodation in Burlington House led to the Chemical Society vacating its former rooms. Although the Royal Astronomical Society continued to occupy the same part of Burlington House as before, it was allocated in addition the second floor of the Chemical Society's former premises. This provided useful extra storage space for the library and an additional small meeting room.

Council decided, however, that a thorough reorganization of the Society's premises was also required. First, the Council Room was restored to a more dignified state than it had been in for many years, combining elegance with the practical advantages of greatly improved lighting, noise insulation and ventilation. It was also decided to cease using the Meeting Room in the autumn of 1969 and, until the Savile Row lecture theatre was ready, to accept the kind offer of the British Academy to hold the Ordinary Meetings of the Society in its lecture theatre in Burlington House, formerly used by the Royal Society. It was agreed that the Society's old Meeting Room should be divided into two parts. It was planned that at the courtyard end there would be two more offices which would not only bring relief to the overcrowded General Office but would also provide a place where the officers could meet without having to close the Grove-Hills Library for that purpose. The larger part, at the rear end, would be made into a Fellows' Room, which could be used for reading, discussion and casual meetings.

These building and decorating plans could not have been envisaged if the Society had not been able to bear the expenses involved in carrying them out. For most of the decade the Society's finances were under the expert control of Gerald Merton. He had originally been appointed Treasurer of the Society on 1956 October 12 and had coped so well with the difficulties caused by inflation that, although Standing Orders required that 'no Fellow shall serve continuously in any office for more than ten years', it was decided by Council to ask him to accept nomination as Treasurer for the session 1966−67, which he agreed to do. He had also been responsible for the agreement made with Blackwell Scientific Publications Ltd concerning the Society's publications and he continued to monitor the agreement through the early stages. In 1967 he was succeeded as Treasurer by Alan Hunter, a former Secretary of the Society, who continued as Treasurer for the rest of the decade and beyond.

At the beginning of the decade the Society benefited from the legal decision that made it possible to reclaim income tax on the subscriptions by Fellows paid under seven-year covenants. Nevertheless, in 1960 there was an overall deficit of nearly £4000 due to the expansion of the Society's work. The main deficit was on the Publications Account, nearly £7000, and arose from the cost of providing free to all Fellows two volumes of *Monthly Notices* during the year instead of the usual one volume, together with the additional cost involved in producing the new *Quarterly Journal*. At the same time volume 3 of the *Geophysical Journal* was rather larger than had originally been planned. To offset this deficit, application was made to the Royal Society for an emergency grant from the Parliamentary Grant-in-Aid of Scientific Publications, and the very welcome sum of £4000 was received. This was a non-recurrent grant, and as the expenditure on printing was unlikely to decrease it was decided by Council in 1961 that Fellows who required *Monthly Notices* or the *Geophysical Journal* should be charged one pound for each volume.

The continual rise in printing costs led in 1965 to a rise in the charge to Fellows purchasing either of these journals to two pounds a volume. Since Fellows were asked to pay at times that depended on the completion of a volume, this system was criticized as less convenient than one in which an annual charge is made that is known in advance. Council therefore decided to revert in 1968 to invoicing annually instead of by the volume and set an annual charge for Fellows of five pounds for *Monthly Notices* and three pounds for *Geophysical Journal*. These charges represented a slight saving to Fellows because, on the advice of the Society's publishers, it had been possible to secure more favourable printing terms than previously.

An added burden on the Society's finances was the payment of rates. Following a change in the law, the Society, for the first time since it came to Burlington House in 1874, was called upon by Westminster City Council to pay rates in the year 1964−65. The amount (£115) was at a reduced rate of one-tenth on a nett rateable value of £3180 but the proportion payable increased year by year until 1968−69 when the full amount less a 50 per cent statutory abatement was charged.

The general state of the Society's finances during the 1960s was satisfactory, despite some fluctuations. Thus a deficit of £6872 in 1963, most of which was on the Publications Account, was followed by a good recovery in 1964 when the overall deficit was only £433. Although this improvement continued in 1965, when there was an overall surplus of £4947, the financial outlook, in view of continued inflation, gave cause for serious concern, since the costs of producing the Society's journals and other expenses were increasing rapidly. However, the results for the following year, which was the first year to include accounts rendered by the Society's publishers, Blackwell Scientific Publications Ltd, of Oxford, were better than expected and there was still a substantial surplus of £3744. This resulted from the increased amount published and accounted for that year and the ability of the Society's publishers in keeping down publication costs. The turnover on the Publications Account was nearly £30000, as compared with £4800 only 10 years before and £700 at the beginning of the century! The arrangement with the publishers proved to be highly beneficial for the Society's finances. In 1967, for the third year in succession, a surplus was reported. It was a record one of £8505 and was made almost entirely on the Publications Account. In 1968 there was again a surplus of comparable amount, although a contribution of £1000 had been made by Council towards the expenses of the IAU General Assembly at Brighton in 1970, and a sum of £4000 was placed to reserve towards the expected costs of the impending accommodation changes. The buoyancy of the Publications Accounts continued in 1969 and made it possible to set aside another £4000 against reorganization costs and still record an overall surplus of £2053. An advantage of a buoyant Publications Account was that the Society became less dependent upon revenue from investments and the Finance Committee was better able to build up the Society's capital.

In 1961 a Special Purposes Fund was created for promoting the study and advancement of astronomy and geophysics and of any new lines of research in which the Society might concern itself in the future, by grants for purposes appropriate either directly or indirectly to that end for which the ordinary revenue of the Society is not normally available. As the result of appeals to Fellows of the Society some generous gifts were received and, in addition, some special allocations were made from surplus income on other funds and from the Percy Morris Bequest of £100 in 1962. Grants from the income were made in connection with conferences and other special meetings, the fund proving particularly useful in paying the travelling expenses of invited speakers and research students to take part in special discussions, thus adding greatly to their usefulness.

In 1960 a special Supplementary Fund was established to increase the income for the George Darwin Lectureship so as to ensure the fee and make possible grants towards the lecturers' expenses. A capital sum of £1000 was provided for this purpose out of the income tax on covenanted annual contributions of Fellows in the years 1956–59 which the Society successfully reclaimed in 1959 and 1961. The fund for the Harold Jeffreys Lectureship was created in 1962.

A regrettable incident concerning geophysics which had a minor adverse financial effect on the Society occurred in 1968. In 1966 February a symposium was held at Newcastle-upon-Tyne on non-elastic processes in the mantle, organized under the auspices of the International Upper Mantle Committee. The RAS Council agreed that the proceedings should be published in the *Geophysical Journal* and a geophysicist who was not a Fellow of the Society was invited by the Geophysical Secretary to collate and edit the proceedings. These were published in volume 14, Nos 1–4 in 1967 October. Arrangements were then made for a hardback edition bearing the title *Non-elastic Processes in the Mantle* to be published the following year. Unfortunately, the editor felt that he had not received adequate recognition of his editorial work and claimed compensation, threatening to sue the Society for damages if his claim was not met. Council took legal advice and, largely because of the vagueness of some of the original arrangements and the possibility of misunderstandings, was advised to seek a settlement. In due course agreement was reached on the payment of £250 to the editor. With the payment of legal costs on both sides, the total expense to the Society was £398.

One of the financial responsibilities of the Society is the maintenance of its library, which dates from the early days of the Society. It contains the most comprehensive collection of astronomical and geophysical works in the Commonwealth and ranks among the great specialist libraries of its kind in the world. It is rich in historical material and contains many early and rare editions of astronomical and mathematical works. As the Society grew many books were presented or acquired when they were published, and the Society obtained by exchange the relevant publications of other learned societies and of observatories. These, together with purchases, eventually gave a virtually complete coverage of the significant literature of astronomy and geophysics, and the library came to be regarded as primarily a reference library of last resort to which workers in the field could turn when the resources of their own institutions were inadequate for their needs. In

1963 following a detailed statement made to it by the Chairman of the Library Committee, G.J. Whitrow, Council decided that despite inflation everything possible should be done to maintain the library so that it could continue to provide this essential service.

In 1964 a generous gift of £1000 was made to the Society by Mr Warin F. Bushell, who had been a Fellow since 1908. He asked that the money should be used for the creation of a permanent fund, the income from which would be available for all purposes, but he expressed a preference for its use for the needs of the library. Council therefore decided that the income from this fund should be allocated to the Library Fund. In the same year a legacy of £200 was also received from the estate of the late Mr A.P. Watson for the repair and binding of books and this too was allocated to the same fund.

In 1966 the library received a very welcome grant of £1400 from the Pilgrim Trust following an appeal by Council, at the request of the Library Committee, for financial assistance to enable it to continue the work of repairing and rebinding the Society's oldest and most valuable books and manuscripts. The Trustees specifically stated that the grant should be spent on the renovation of pre-1800 books, manuscripts, letters, etc. In thanking the Trust for this generous response to the Society's appeal, the President referred to the personal interest that had been taken in the matter by two of the Trustees, Lord Bridges and Lord Crawford (the latter's grandfather had been President of the Society from 1878 to 1880, but since he had not then succeeded to the Earldom, he was known at the time as Lord Lindsay), both of whom had visited the Society and its library.

During the course of the decade there was increasing awareness by the Society of its obligation to conserve all the valuable material in its possession. Evidence that this concern was not restricted to books and periodicals was revealed in connection with the celebrations in 1970 of the 150th anniversary of the foundation of the Society, when an exhibition of various items of historical interest belonging to the Society which are not normally on view was displayed in the Main Library in special cases that had been acquired for use on this and subsequent occasions. It was gratifying to discover how much material relating to the foundation and early years of the Society had survived, although the archives of the Society had never been catalogued or even roughly sorted! Moreover, it was known that some material that was formerly in the Society's possession had been lost, in particular the copy of Francis Baily's *Journal of a Tour in Unsettled Parts of North America in 1796 and 1797* which had inserted in it various maps, letters, etc., by Baily, who was one of the leading figures in the early days of the Society. An even more serious loss was the set of Minute Books of the Spitalfields Mathematical Society, founded in 1717 and absorbed into our Society in 1846. Unfortunately, they had been passed on to the London Mathematical Society some time in the 1930s. The library of that Society was housed in the Library of University College, London, and these Minute Books appear to have vanished irretrievably in the severe bombing of that college during the Second World War.

In the summer of 1969 search for a letter known to have been written to the Society by Michael Faraday led to the discovery of other interesting and early correspondence, the whereabouts of which had not been known for many years, including the letter from the Duke of Somerset explaining that he felt obliged to decline the Presidency of the Society on its foundation in deference to the opposition of the then President of the Royal Society (Sir Joseph Banks) to the formation of the Astronomical Society, and the subsequent letter from Sir John Herschel declining on behalf of his octogenarian father the offer of the Presidency when first made to him. Also, after long search, a trunk of Herschel material was found in the basement of the Society's apartments, buried under a pile of old periodicals and mailbags. Similarly, in 1968 the portrait of Joseph Middleton, who founded the Spitalfields Mathematical Society in 1717, was found in the basement in a very neglected condition. This portrait was subsequently cleaned and varnished. Incidentally, this was not the only portrait in the Society's possession to be cleaned during the decade. The portrait of John Goodricke (1764–1786) which was in the Meeting Room was found in 1966 to have been attacked slightly by fungus. Council agreed that this brilliant pastel should be restored by the fine art restorer Arthur B. Drescher, since it was regarded by experts as of finer quality than any portrait of similar type in the National Portrait Gallery.

An important question that arose early in the decade in connection with the library concerned the adequacy of the Society's insurance arrangements, since it depends mainly on the estimated financial value of the contents of the library. In 1964 this was roughly calculated to be not less than £100000, whereas the total sum for which the Society was then insured was only £39000, an amount that had been agreed upon in 1939. Council therefore decided to revise its insurance arrangements to take account of this revaluation of its property.

From 1964 the sale and loan of astronomical prints and slides by the Society was administered by the library staff. The collection of slides and prints was overhauled and many additions made. A new catalogue was prepared, much of the work on it being done by the secretary of the Photographic and Instruments Committee, R.H. Garstang. In 1966 the library extended its service of slides and prints by arranging for the sale of colour slides from the Mount Wilson and Palomar Observatories. This proved to be a highly successful venture, some 4000 colour slides being sold in the first seven months, and a useful contribution was thereby made towards the cost of the library services.

The successful management of the Society during the 1960s, with the continuing expansion of its membership and activities, would have been impossible without the loyal and efficient services of its senior staff. By 1963 the growth of the Society and particularly of its publishing activities made it necessary to increase the staff and reorganize the office. E.C. Rubidge, who had been in charge of the office with the title of Assistant Secretary since 1956, was given the title of General Secretary, and Mrs M.M. Bassett, who had shouldered much of the burden of administrative work, became the new Assistant Secretary. In 1965 the Society's Librarian, Mrs B. Harrington, resigned, after more than five years with the Society, as a consequence of her move away from London. Her resignation was accepted with regret but the Society was fortunate in being able to appoint as Librarian Dr R.E.W. Maddison, formerly the director of a small industrial chemical laboratory and an expert linguist with a professional interest in the history of science. (It was due to him that the letters and Herschel material mentioned above were successfully located.) For the rest of the decade the senior staff remained unchanged and the Society continued to depend heavily on their valuable and devoted services.

5. Meetings of the Society

During this decade Ordinary Meetings of the Society continued to take place on the second Friday of the month. In 1960 eight were held in Burlington House and also three Geophysical Discussions. In 1961 there were the same number of Ordinary Meetings and four Geophysical Discussions, but only seven of the former were at Burlington House. This was because it was decided to devote the December meeting to the subject of the British contribution to space research and it was expected that there would be a larger attendance than usual, so that more extensive premises would be required to provide adequate accommodation. Through the courtesy of the Council of the Institution of Civil Engineers the meeting was held in the Institution's lecture theatre at Westminster, and the attendance fully justified the arrangement. Those who spoke included Sir Harrie Massey, Chairman of the British Committee for Space Research, and S.F. Follett, Deputy Director of the Royal Aircraft Establishment, Farnborough. Among the topics discussed were rocket studies of stellar ultraviolet radiation, far ultraviolet studies of the solar corona and chromosphere, the possibility of lunar and planetary observations from space vehicles and radio astronomy from artificial satellites.

On 1961 May 24 at a luncheon at St John's College, Cambridge, which was arranged by the Department of Geodesy and Geophysics, a suitably bound volume of *The Earth Today* (volume 4 of the *Geophysical Journal*) which had been produced in honour of Sir Harold

Jeffreys's seventieth birthday was presented to him by the President of the Society, Professor McCrea. On July 14 a visit by Fellows and their guests to the Royal Greenwich Observatory at Herstmonceux Castle took place, at the kind invitation of the Astronomer Royal. The party numbered about 90. On October 19 a number of Fellows had the opportunity to meet a delegation from the Academia Sinica of Peking before a lecture given at the London Planetarium on 'Astronomy in Classical China' by Dr Joseph Needham.

There were nine Ordinary Meetings and five Geophysical Discussions in 1962, eight of the former being at Burlington House. The additional meeting took place on September 6 in the Physics Lecture Theatre of the Queen's University, Belfast. At the meeting on November 9 the Society commemorated the bicentenary of James Bradley (1693–1762). The President, Professor McCrea, spoke on the significance of Bradley's discovery of aberration. Other speakers were D.E. Blackwell of Oxford, where like him Bradley was Savilian Professor of Astronomy, Sir Richard Woolley, the eleventh Astronomer Royal, who spoke on Bradley's work as the third Astronomer Royal, H.R. Calvert of the Science Museum, London, who showed the model that Bradley devised to explain the principle of the aberration of light, and P.S. Laurie of the Royal Greenwich Observatory, who displayed photographs of Bradley's instruments.

The same number of Ordinary Meetings (nine) and Geophysical Discussions (five) took place in 1963 as in the previous year, one of the former being held on September 20 in the rooms of the Royal Society of Edinburgh as part of the programme arranged for the visit by the Society to Edinburgh. An additional all-day meeting and discussion of interest jointly to astronomers and geophysicists took place in the Society's premises at Burlington House on January 18. The subject was 'The origin and constitution of the planets'. It consisted of three sessions on the thermal state and history of the Earth, the mechanical properties and constitution of the planets, and the origin of the planets by accretion, respectively. Advantage was taken of the presence in this country of H.W. Menard, H. Suess and H.C. Urey, all of whom accepted invitations to read papers.

Eight Ordinary Meetings were held in Burlington House in 1964, another took place in the Physics Building of the University of Newcastle-upon-Tyne on July 24, and there were three Geophysical Discussions during the year. At its meeting on February 14, following the President's Address, the Society marked the quatercentenary of the birth of Galileo Galilei with an address by G.J. Whitrow on 'Galileo's significance in the history of astronomy'. Following the General Assembly of the International Astronomical Union in Hamburg, the opportunity was taken to invite a number of astronomers from abroad to attend an additional meeting in the Society's rooms on September 7 and 8 devoted to a discussion on 'Stellar evolution'. Among those who took part were P. Demarque, A.J. Deutsch, A. Przybylski, A. Sandage, A.D. Thackeray and B. Westerlund. On January 24 an all-day Geophysical Discussion was held on 'The influence of the Earth's magnetic field on ionospheric and atmospheric phenomena'.

In 1965 although the number of Ordinary Meetings (eight) was the same as in the preceding year, there were no fewer than five Geophysical Meetings, an indication of the growing interest and activity in that field. The summer meeting of the Society was held at Dublin on September 7 at the invitation of the Royal Irish Academy. Since it was the centenary year of the death of the great Irish mathematician Sir William Rowan Hamilton, a joint symposium was also held with the Royal Irish Academy on Hamilton's work. Other features of a three-day programme included a colloquium on 'High-energy particles in astronomy', a Geophysical Discussion on 'Theory of the Ice Ages' and a public lecture by Father D.J. O'Connell on 'Binary Stars'. The most remarkable feature of this highly successful out-of-town meeting of the Society was that it was honoured by the presence of the President of the Irish Republic, Eamonn de Valera, who attended many of the scientific discussions.

Again, in 1966, there were eight Ordinary Meetings and five Geophysical Meetings, including an all-day Discussion on 'Coral Clocks'. These were all held in the Society's rooms at Burlington House. An additional meeting of the Society was held at Exeter, by invitation of the University, on September 14. As in 1965, the complete programme of this out-of-town meeting was spread over three days and included a colloquium on some historical aspects of astronomy, geophysical discussions on the solar wind and solar effects on the upper atmosphere, and a visit to the Norman Lockyer Observatory at Sidmouth, which had been taken over recently by the Physics Department of the University of Exeter.

Nine Ordinary Meetings and three Geophysical Meetings were held in the Society's rooms at Burlington House in the course of 1967. The Society also participated in a joint meeting with the Physical Society and the Institute of Physics held at Jodrell Bank from September 5 to September 7. The subject was 'The Physics of Quasars'. No summer meeting took place that year, however. These meetings had always been held at centres well away from London in response to local initiative, usually taken by a university. Council felt that the continued success of such meetings must depend crucially on local enthusiasm and it decided that, although it was always prepared to consider spontaneous invitations for the Society to meet in this way, it would be a mistaken policy to stimulate such invitations.

In 1968 there were eight Ordinary Meetings and five Geophysical Discussions. All were held at Burlington House except for the Anniversary Meeting of February 8 and the Ordinary Meeting of October 11. They were held in the Wellcome Lecture Hall of the Royal Society's new premises at 6 Carlton House Terrace, by kind permission of the President and Council of that Society. On 1968 May 9 a special meeting and discussion took place on the subject of the education of astronomers. On July 18 a suitably bound volume of *The Earth in Space*, the hardback edition of volume 15, Nos 1 and 2, of the *Geophysical Journal* comprising a series of papers contributed by colleagues and students of Professor Sydney Chapman in honour of his eightieth birthday, was presented to him in London by the President on behalf of the Society. Afterwards a number of contributors to the volume joined members of the Council and other representatives of the Society at a luncheon given in his honour at the Café Royal.

Eight Ordinary Meetings and three Geophysical Discussions were held in 1969. An extra meeting, on October 1, was held jointly with the Royal Institute of Navigation in the rooms of the Royal Geographical Society, the subject being an illustrated lecture by J.P. Mayer and L.G. Richards of NASA on 'Space Navigation — Apollo and Beyond'.

Eight ordinary meetings and some associated early afternoon meetings on both astronomy and geophysics were held in 1970. There was also a joint meeting with the Institute of Physics and the Physical Society on space physics, aeronomy and geomagnetism. Council accepted the proposal by P.C. Kendall, H. Rishbeth and V.C.A. Ferraro that it should support future meetings by scientists working in these fields. They soon became known as Magnetosphere, Ionosphere and Solar-Terrestrial Relations (MIST) meetings.

The decade closed with the Sesquicentennial celebrations of the Society. The Anniversary Meeting on 1970 February 13 was held in the Meeting Room of the British Academy at Burlington House, and after the President's Address, the Society commemorated the Sesquicentenary of its foundation in 1820. Four Fellows addressed the Society on various aspects of its early history. G.J. Whitrow spoke on 'Some prominent personalities and events in the early history of the Royal Astronomical Society', referring to Francis Baily, Henry Colebroke, William Pearson and John Herschel. H.R. Calvert spoke on 'The presentation of Caroline Herschel's telescope to the Society by Sir John Herschel', Herbert Dingle spoke on 'Sidelights on astronomy during the Society's history', and P.S. Laurie spoke on 'The Society and the year of revolutions', the latter being 1848 and the subject the Society's entry into the diplomatic field to ensure the continued publication of the important astronomical periodical, *Astronomische Nachrichten*, following the invasion of Schleswig-Holstein by Prussia, the editor being

H.C. Schumacher (1780−1850), director of the Altona Observatory in Holstein, which was then in Denmark.

The formal proceedings of 1970 February 13 culminated in a Celebration Dinner at the Dorchester Hotel, Park Lane, attended by numerous Fellows and their guests and specially invited representatives of national and international organizations. The Society's guest of honour was Professor Otto Heckmann, President of the International Astronomical Union. Among those present were Miss Caroline Herschel, great-great-granddaughter of the Society's first President (and the sole remaining person to bear the family name), and her sister Mrs. Shorland. The toast of the Society was proposed in a felicitous speech by Professor Heckmann, in the course of which he pointed out that 'When, 150 years ago, the astronomers of Great Britain and the continental countries had reached nearly the same level of maturity, it was not elsewhere, it was here, that a society was created with almost the character of an academy in order to cultivate astronomy in particular. Right from the beginning the tasks of the Society were defined in such a way that, notwithstanding its national individuality, the international character of astronomy was duly acknowledged.' He concluded that it would be almost impossible to imitate the RAS anywhere abroad, because 'the basic conditions in the country which created the "Mother of Parliaments" are too specific.'

The Sesquicentennial celebrations continued during the year. A commemorative stamp was issued by the Post Office on April 1. A special series of Celebration Lectures were delivered at the Royal Institution on 1970 June 26. After introductory remarks by the President, Sir Bernard Lovell, a lecture on the Earth by Sydney Chapman was read by T.G. Cowling. (Sydney Chapman had died on June 16 after a short illness, and Cowling paid a tribute to him before reading the talk that Chapman had prepared.) D.E. Blackwell then spoke on 'The Sun and the Interplanetary Medium'. Sir Richard Woolley, Sir Martin Ryle and H. Bondi, respectively, spoke on 'The Stars and the Structure of the Galaxy', 'Energy Problems in Extragalactic Research', and 'Astronomy of the Future'.

These Celebration Lectures were followed in the evening by a reception given in the Society's honour by the Lord Mayor and Corporation of the City of London at Guildhall. (It will be recalled that for the first decade of its existence the Society was called the Astronomical Society of London.) Some 600 Fellows and their guests joined about 300 of the City's own guests in celebrating the Society's Sesquicentenary. There were no speeches, but public announcement was made of the following exchange of telegrams between the President and Her Majesty the Queen:

Plate 12. The Sesquicentenary Exhibition in the Society's library 1970. (© Royal Astronomical Society.)

On the occasion of the 150th anniversary of the foundation of the Royal Astronomical Society the Fellows of the Society assembled in celebration at Guildhall as guests of the Corporation of London send their warmest and most loyal greetings to Your Majesty, their Patron.

BERNARD LOVELL
President

Please convey to all Fellows of the Royal Astronomical Society my sincere thanks for their kind and loyal message sent on the occasion of the celebration of the 150th Anniversary of the foundation of the Society which as Patron I much appreciate. I send to you all my warmest congratulations and my best wishes for the future.

ELIZABETH R

Throughout the period from 1970 February to September a special exhibition of items of historical interest in the Society's possession, as mentioned above, was held in the Main Library. These included the Society's Charter, the Obligation Book containing the signatures of Royal Patrons of the Society, the letter from the Duke of Somerset explaining why he felt obliged to decline the Presidency of the Society on its foundation, and the subsequent letter from John Herschel declining on behalf of his father the offer of the Presidency when first made to him. Plate 12 is a photograph of the exhibition.

The exhibition remained open until September to enable overseas Fellows who attended the XIVth General Assembly of the International Astronomical Union in Brighton in August to visit the Society's premises and see the treasures. Although the IAU meeting was not the responsibility of the Society, many members including the President were closely involved in the organization of what was a most successful meeting.

A memorable year and decade came to a close when the President entertained officers and members of Council and their wives to dinner at the Royal Society's premises in Carlton House Terrace on 1970 December 10.

CHAPTER 6
THE DECADE 1971−80

R.J. TAYLER

The Scientific Background

The first bye-law of the Society states that its object is the encouragement and promotion of astronomy and geophysics and, before turning to a discussion of the domestic affairs of the Society, it seems appropriate to comment on how the two sciences developed during the decade.

In astronomy it was a decade of solid progress and introduction of new techniques, even though the individual discoveries did not appear so dramatic as those of quasars, the microwave background and pulsars in the previous decade. Pulsars and quasars continued to be enigmatic objects. Most authors accepted that quasars were at cosmological distances and that their nuclear activity probably involved a massive black hole but a detailed model of the central object was still lacking. The discovery of twin quasars, almost certainly produced by gravitational lensing, was dramatic. The model of the pulsar as a rotating neutron star was well established and the properties of the binary pulsar gave further support for the general theory of relativity but precisely why and how they pulse remained a mystery. There was a general recognition that quasars, radio galaxies and active galaxies were all manifestations of the same general phenomenon of nuclear activity probably associated with the accumulation of mass in galactic centres.

The good qualitative agreement of observations of the microwave radiation and the abundances of light elements with the hot big bang cosmological theory led to increased interest in the departures from strict uniformity; angular variations in the microwave radiation and the observed clustering of galaxies. Inside galaxies the understanding of the interstellar medium was transformed with the discovery of dense molecular clouds and of very hot interstellar gas emitting ultraviolet radiation and X-rays. The study of the rotation curves of galaxies led to the realization that their mass distribution must be different from their light distribution while many elliptical galaxies were discovered to be triaxial rather than spheroidal.

Study of stars concentrated on close binary stars with the discovery that many of the most violent stellar phenomena such as novae and X-ray emission were associated with mass exchange between components. In the case of the Sun, the solar neutrino problem was a puzzle throughout the decade, the new subject of solar seismology promised to give information about the otherwise hidden properties of the solar interior and very detailed studies were made of the dynamics and magnetohydrodynamics of its outer layers.

Many of these discoveries were made as a result of instrumental developments. The United Kingdom astronomer gained the use of such new ground-based telescopes as the Anglo Australian Telescope, the United Kingdom Schmidt Telescope, the United Kingdom Infrared Telescope and the Cambridge 5km Radio Telescope, while the new observatory on La Palma, the MERLIN array at Jodrell Bank and the United Kingdom Millimetre Telescope were under active development. With the new telescopes came much-improved detectors. The new facilities were reflected in the contents of the scientific meetings of the Society and of its journals. It was, however, perhaps in space astronomy that the most dramatic development occurred. X-ray astronomy was the largest growth topic in the journals, with the United Kingdom Ariel 5 satellite playing an important role and ultraviolet astronomy was transformed by the success of the International Ultraviolet Explorer. Although emphasis has been given above to the United Kingdom's facilities, astronomy is an international science and the Society has an international Fellowship and results reported to the Society were obtained using all of the leading facilities in the world.

Members of the Society were prominent in many of the discoveries of the decade and they received many honours and awards for their work. Of particular note was the award of the Nobel Prize for Physics for 1974 to Professors Sir Martin Ryle and Anthony Hewish, both of the Mullard Radio Astronomy Observatory, Cambridge. The award was particularly related to Ryle's fundamental work on the design of radio telescopes and to the results obtained with them and to Hewish's role in the discovery of pulsars.

The developments in geophysics and planetary science were also very important and many of them also benefited from the development of space techniques. In geophysics the most dramatic change was the general acceptance and development of the new science of plate tectonics. Continental drift was the new orthodoxy, with precise measurements of the movement of the plates becoming possible and with a great impetus being given to studies of the phenomena at plate boundaries. There were also important developments in the study of magnetohydrodynamic dynamos, this being stimulated by the discovery of additional magnetic fields in the Solar System. The increase in computational power and new mathematical techniques enabled inversion of seismic data to provide much more detailed information about the Earth's interior. Satellites provided a wealth of experimental data about the Earth's magnetosphere and about solar–terrestrial relationships in general and longer distance satellites, such as the Voyager spacecraft, provided new insights into the structure of the outer planets and satellites of the Solar System. It is in the context of these developments in our sciences that we now turn to what is more strictly the history of the Society.

General Introduction

If one word can be said to characterize the activities of the Society in the period from 1970 to 1980, it is 'growth'. The previous chapter ended with the Society celebrating the 150th anniversary of its foundation, with the largest membership that it had ever possessed, with its premises newly organized for the better administration of the Society and the convenience of the Fellowship and the staff, and with scientific activities at a very high level. During the 1970s there was a substantial growth in the number of hours devoted to scientific meetings, in the size of the Society's journals and in the turnover of the Society's accounts. Although the membership of the Society continued the rapid growth of the 1960s for the first part of the decade, it remained fairly stationary after that, with maximum membership of 2812 occurring in 1975. The large increase in financial turnover was substantially due to inflation and it was accompanied by an unprecedented increase in the Annual Contribution of Fellows. The levelling off and reduction in membership was certainly partly caused by this but another important influence was the revision of the bye-laws in 1977, which led to the removal of members who were in default in the December of any year. As a result, starting in 1977, the number of members listed in the table of progress accompanying the *Annual Report of Council* was the actual number who had paid their Annual Contribution in the year or who were Associates or entitled to free Fellowship. As a consequence, the 1979 membership of 2742 probably is the largest ever true membership of the Society.

One minor change in membership of the Society was determined in 1977 when it was agreed that Institutional Membership of the Society should be abolished at the end of 1978. Since 1924 it had been possible for institutions to join the Society for the same subscription as Fellows. They were able to obtain journals on the same conditions as Fellows and to send a representative to Ordinary Meetings. Although in 1977 there were only five such members, Council decided that it was un-reasonable for institutions to obtain the Society's journals for the same price as Fellows by simply paying what was, in effect, a nominal membership fee.

Finance

An important development in the decade was the realization that the Society needed to give careful consideration to its financial estimates every year. In the immediate past, apart from occasional financial crises, the Annual Contribution of Fellows and the prices for publications were changed when the annual accounts in the Spring showed that this was necessary. This was a leisurely process; publication prices could not be changed for nine months when the next year's subscriptions were due and an increase in Annual Contribution could take even longer because it required a change in the bye-laws. This procedure was totally inadequate to deal with the inflation of the 1970s.

In the spring of 1972, Council and then the Annual General Meeting learnt that there had been a deficit on the Society's accounts in 1971 of more than £7000, which represented about one-eighth of the Society's financial reserves. The author of this chapter was attending his first Annual General Meeting as Secretary when these results were announced and at that time it would have been difficult to foresee the very healthy financial state of the Society at the end of the decade. A consequence of this deficit was a series of meetings of the Society's Finance Committee which led to the proposal, accepted by the Annual General Meeting in 1973, that the Annual Contribution should be doubled from £5 to £10 in 1974; in fact a new category of Fellows aged under 30 was introduced and they paid £7, while Junior Members paid £3.50. The 1971 deficit also led to the realization that there had to be careful advance estimates of the turnover and profitability of the Publications Account and of the cost of the Society's other activities. Soon afterwards this was followed by the introduction of a July Council Meeting at which the following year's budget was the principal item of business.

Also following the 1971 deficit, the Treasurer, A. Hunter, put in an application to the Royal Society for a grant towards the payment of rates on the Society's premises and a grant of £1206 was received in 1973 February. At that time officers and Council were under the mistaken assumption that rating grants could only be obtained when the Society's finances were in poor condition. Later in the decade, Hunter's successor, W.H. McCrea, argued that a request for a rating grant should be made every time that the Royal Society announced that such grants would be available and, in fact, a grant was obtained every year from 1978 onwards.

The Society was fortunate that it entered the decade with its journals in a financially sound state. They were also able to publish an increasing amount of good science in a rapid and efficient manner. For some years the journals had been earning a surplus which had balanced out fluctuations in the other income of the Society and which had also added to the Society's reserves. This was in healthy contrast to some earlier times when the Society had required financial aid for the publication of its journals. The good financial position was obviously influenced by the great growth in higher education that had recently taken place, not only in the United Kingdom but throughout the developed world, and by the increasing opportunities for research workers both in ground-based and in space astronomy and also in various branches of geophysics. The opening of new universities and other institutions was a stimulus to the library circulation of the journals. It was, however, recognized that there was a risk if the general activities of the Society were to be too dependent on the surplus from publications. In 1971 the income from members was only one-seventh of the turnover of the Publications Account. If the general running of the Society was to be largely independent of the journals, there was need for a continuing increase in the Annual Contribution of Fellows, which rose to £15 in 1976, £18 in 1978 and £24 in 1980. By the end of the decade the general running of the Society was essentially independent of the journals, although the income from members was still only one-seventh of the publications turnover because of a large increase in journal size. The satisfactory financial position was helped considerably by a rise in investment income from £3000 in 1970 to £31 000 in 1980.

On various occasions Council was asked to consider the introduction of new types of Fellowship carrying a reduced Annual Contribution. Such requests were received from Fellows overseas or even at a large distance from London within the United Kingdom, who could rarely attend meetings of the Society, from Fellows who were temporarily unemployed and from Fellows who were required to retire at the age of 60, five years before they were eligible to claim free Fellowship. There were also problems relating to overseas Fellows from countries with rigorous exchange controls. Although officers and Council were sympathetic to many of these pleas, they felt that it was necessary to recognize that the loss of income from the introduction of special classes of membership, accompanied as it must be by a further increase in the regular Annual Contribution, would be much greater than the loss due to the resignation of a small number of members. The figures for membership, which have already been quoted, suggest that their decisions were correct.

The journals did continue to make a good surplus, since in a time of high inflation cautious budgeting almost certainly implied such a surplus provided that the library circulation of the journals held up. In fact the circulation held up very well throughout the decade, although at the very end cuts in library budgets were beginning to be of concern. One factor which always produced uncertainty in the results of the Publi-

Table 2

Year	1970	1971	1972	1973	1974	1975	1976	1977	1978	1979	1980
Annual Contribution	£4.20	£4.20	£5.00	£5.00	£10.00	£10.00	£15.00	£15.00	£18.00	£18.00	£24.00
Income from Fellows	£10K	£10K	£12K	£12K	£21K	£24K	£31K	£33K	£39K	£39K	£51K
Investment income	£3K	£3K	£2K	£3K	£4K	£6K	£7K	£12K	£14K	£22K	£31K
Administration and General Account turnover	£27K	£30K	£32K	£30K	£44K	£55K	£66K	£74K	£83K	£99K	£112K
Administration and General Account surplus	−£8K	−£13K	−£9K	−£7K	−£6K	−£8K	−£6K	−£4K	−£2K	−£3K	+£14K
Publications Account turnover	£76K	£71K	£121K	£98K	£228K	£200K	£235K	£242K	£268K	£278K	£344K
Publications Account surplus	£14K	£6K	£11K	£13K	£45K	£19K	£42K	£37K	£45K	£28K	£34K
Investments book value	£61K	£61K	£58K	£58K	£52K	£85K	£96K	£124K	£169K	£190K	£245K
Investments market value	£64K	£82K	£91K	£66K	£37K	£101K	£111K	£173K	£197K	£200K	£264K

cations Account was the relation of the dollar to the pound. American subscribers liked to have a price quoted in dollars. This raised no problems when currencies had fixed parities but, when these were abolished, there was a strong likelihood that the parity would change after the journal price was fixed. When the pound fell relative to the dollar, this led to a substantial additional income for the Society.

Table 2 summarizes the decade's financial details. One particular point to notice is the extreme volatility of the market value of the Society's investments in the early part of the decade when Council was trying to formulate a sound financial policy. Two further special points should be mentioned. The exceptionally large turnover of the Publications Account in 1974 arose because of a change in bringing volumes of the journal to account; well over one year's publications were included in the accounts that year. For most of the decade progress was made in eliminating the deficit on the Administration and General Account. It was expected to record a small surplus in 1980 but the surplus was very much larger than planned mainly because the Society was without an Executive Secretary for much of the year and this led to reduction of other expediture.

Publications

At the start of the decade both *Geophysical Journal* and *Monthly Notices of the Royal Astronomical Society* were at a stage of rapid growth and *Quarterly Journal* was well established. The first two journals were looked after by Editorial Boards, which had been set up at the beginning of 1970 to assist the Secretaries, who according to the bye-laws of the Society were still required to be the Editors of the Society's journals. The Editorial Boards replaced papers' committees which had assisted the Secretaries in the recent past. Previously all final decisions about acceptance and rejection of papers were taken by Council, advised by papers' committees. Now the Editorial Boards had the whole responsibility for the contents of the journals except that they were unable to reject papers. All cases for rejection still had to be brought to Council; the large amount of Council time, which might have been spent in considering marginal papers, was minimized by the success of the Editorial Boards in persuading many authors to withdraw papers voluntarily.

Both *Geophysical Journal* and *Monthly Notices* were published irregularly with an issue being made up when enough material was available to produce a copy of about a prearranged size. This had the advantage that publication delays could be minimized but it also meant that the number of issues or volumes to be published in any year was not known in advance. This meant that subscribers could only be invoiced for the estimated number of volumes for a year, with the result that they might receive a supplementary invoice at a later date. This was a problem for librarians trying to plan their expenditure. The irregular publication schedule also raised problems for the Society's printers.

In 1972 Council discussed the future of the journals, including the possibility that they might progress to a regular monthly publication schedule and, after discussion with the publishers, Blackwell Scientific Publications Ltd, it was agreed in 1973 that this should start in 1974. It was this change which led to the very large Publications Account turnover in 1974 which has been mentioned earlier because all of 1974's journals as well as some from 1973 were brought to account. As a result of the new schedule, *Monthly Notices* became a genuine monthly journal for the first time; earlier when it had been published monthly there were only issues corresponding to the meeting session. There were some fears about the change to monthly publication because it meant that the size of the journal for a given year had to be estimated when the price was fixed in the middle of the previous year. In fact these estimates were quite successful until the very end of the decade when the continuing growth in size of *Monthly Notices* did lead to increased publication delays because the papers could not be published immediately that they were available without leading to financial losses.

There were many more detailed changes in the arrangements for the publication of the journals. During 1974 it was agreed that papers for the Society's journals need not be communicated by a Fellow of the Society. The main reason behind this decision was that, if the Society's journals were to be successful in international terms, they had to be open to all astronomers and geophysicists who had good papers to submit. *Geophysical Journal* had already established an American Office with two American Editors who handled papers from North America and this venture was successful throughout the decade. 1971 saw the introduction of *Short Communications* in *Monthly Notices*. These were short, urgent papers which were published much more rapidly than ordinary papers. In 1974, it was decided that copies of *Geophysical Journal* and *Monthly Notices* should be sent by air to North America and then distributed by surface mail. This service was later extended to the Far East and Australasia.

The publication of *Memoirs* raised various problems. It had previously been amalgamated with *Monthly Notices* as far as subscriptions were concerned because it did not appear to have a viable future as an independent journal. Council remained concerned that *Memoirs* papers were expensive to produce because they contained a large amount of tabular and diagrammatic material and that they only appealed to a limited number of readers. In 1974 it was decided that authors should pay a page charge to publish in *Memoirs* but this was not a successful move. Authors in the United Kingdom could not, in general, obtain funds to pay page charges and Council was continually receiving requests for partial or total remission of page charges. In fact it was able to use an unexpected grant from the Radcliffe Trustees to meet some of these requests. This was not going to be a successful solution in the long run and in 1977 Council decided that *Memoirs* should cease publication at the end of volume 85. In future, astronomical papers with a large amount of tabular and diagrammatic material would appear on microfiche and would be distributed with *Monthly Notices* which would contain an extended summary of the paper. At the same time page charges were abolished. The first microfiche appeared in 1978 January. The microfiche decision was not very popular and it gradually became established that papers would appear partly in conventional form and partly on 'fiche. The last edition of *Memoirs* was a notable one, being the catalogue of the Society's Herschel archive produced by Dr J.A. Bennett who had spent two years as the Society's archivist.

The policy of *Monthly Notices* continued to be to publish only original research articles on all branches of astronomy and some related sciences. In contrast *Geophysical Journal* occasionally devoted issues or whole volumes to the proceedings of conferences and some of these were also issued as hard cover books. In 1979 April Council agreed that one of the 1981 issues of *Geophysical Journal* should be devoted to a celebration of the 90th birthday of Sir Harold Jeffreys. This followed a good precedent as volume 4 of the journal in 1961 had celebrated his 70th birthday and the Harold Jeffreys Lectureship had been financed as a result of special hardback sales of that volume.

From its introduction in 1960 until 1975 *Quarterly Journal* had been distributed free both to Fellows and to library subscribers to *Monthly Notices*. This meant that almost the whole price of the journal was charged to the Administration and General Account and therefore, in effect, to the Annual Contributions of members. This did not seem sensible in a time of rapidly increasing costs and in 1975 it was decided that the library subscribers should in future be charged for *Quarterly Journal* if they wished to continue to receive it. Most of them did and the subscription price was fixed to take account of the part of the journal which was devoted to house news. This was so successful that by 1980 two-thirds of the expenditure on the journal was being met by outside subscribers.

In 1975 the Publications Committee had long discussions both with our own existing publishers and with a possible alternative publisher. In addition there was correspondence with a further publisher. The latter discussions were not initiated because of dissatisfaction with Blackwell Scientific Publications Ltd but because it was felt that the time had come to review arrangements which had been made ten years before.

The result of these and other investigations was to suggest that the service provided by Blackwell Scientific Publications was competitive. One tangible result of the discussion with our publishers was their agreement to make quarterly payments to the Society based on an estimate of the end-of-year surplus of the Publications Account. This enabled the Society to have use of this money for part of the year and was one factor leading to the increase in investment income.

In 1976 Council introduced a new subscription rate for non-library purchasers of *Geophysical Journal* and *Monthly Notices* who were not Fellows of the Society. Also in 1976 there was a change of printers for *Geophysical Journal* from C.F. Hodgson and Son Ltd of London to Adlard and Son Ltd of Dorking, who already printed *Monthly Notices*. This change was accompanied by the introduction of IBM typesetting for *Geophysical Journal* and this was also implemented for *Monthly Notices* in 1977.

In 1979 Council considered the consequences of the change in bye-laws which no longer required the Secretaries to edit the Society's journals, even in a formal sense. As a result Managing Editors were appointed for the three journals: A.H. Cook for *Geophysical Journal*, R.J. Tayler for *Monthly Notices* and S. Mitton for *Quarterly Journal*. In fact this did not mark any immediate change as Cook had been an Editor of *Geophysical Journal* since it was founded, Tayler had been in effect Managing Editor of *Monthly Notices* since he became Senior Secretary in 1972 and Mitton had edited *Quarterly Journal* on behalf of the Secretaries and Council, since 1976. At the same time it was agreed that Editorial Boards could reject papers without bringing their recommendations to Council provided that each rejection involved the Managing Editor and at least one other Editor. At the same time the right of Fellows to appeal to Council if they thought a paper was being rejected without justification was retained. Council now spent virtually no time discussing the contents of the publications, although Editors still occasionally sought the advice of Council in difficult cases. Because Managing Editors need not be members of Council, it was agreed that they should attend Council once a year to present reports on journals.

All of the Society's journals ended the decade with good circulations and with good international reputations.

Awards and Named Lectures

An important aspect of the activity of Council and through it of the Society, during most of the Society's history, has been the award of medals, the election of Associates and the selection of scientists to deliver the Society's named lectures. In 1971 Council proposed the institution of two additional medals to be named the Halley and Herschel Medals. These were to be awarded for single, specific pieces of research, or related groups of research, in the same manner as the Eddington Medal, with the Halley Medal covering geophysics and the Herschel Medal the parts of astronomy not covered by the Eddington Medal. For some time the provisions concerning fields of research relating to the award of the Eddington Medal had been interpreted somewhat freely but this was rightly felt to be unsatisfactory. The proposal was put to the Annual General Meeting on 1971 March 12 that all three medals should be available for award each year, but, following a speech by Dr D. Lynden-Bell, who argued that there would not be enough really well-qualified recipients for the medals, the proposal was rejected by the meeting. This was the second time in eight years that a Council proposal relating to medals was rejected, the previous one in 1963 being that there should be two Gold Medals, one each for astronomy and geophysics, respectively. In the intervening years there had usually been one Gold Medal in each of these fields but this was a result of Council Standing Orders rather than of a change in the bye-laws.

Following the rejection of the proposal for additional medals, Council discussed the possibility that there should be the three medals in addition to the Gold Medals but that they should each be awarded not more often than once every three years. Later in 1971, Professor V.C.A. Ferraro proposed to Council that the Society should institute a Chapman Lecture in memory of Professor Sydney Chapman, who had died in the previous year. After discussion Council decided to recommend the institution of a Chapman Medal rather than a Halley Medal for geophysics and planetary science. The Annual General Meeting in 1972 agreed to the founding of the Chapman and Herschel Medals with the proviso that they and the Eddington Medal should be awarded not more frequently than triennially. This decision had financial implications. The Eddington Medal had been financed by money raised as a result of a special appeal but the Chapman and Herschel Medals became a charge on the Society's general funds. The first Chapman Medal was awarded in 1973 to D.H. Matthews and F.J. Vine for 'their outstanding contributions to the concept of the spreading of the ocean floor from a single line' and the first Herschel Medal in 1972 to J.P. Wild for 'his outstanding contribution to solar radio astronomy with particular reference to the major technical, observational and theoretical advances which he had made and which have led to the present detailed understanding of the character of the active Sun.'

In 1972 Council made a further change in its policy with regard to the consideration of the award of medals. From very early in the Society's history the award of medals had been regarded as the major item of Council's business and, in fact, the bye-laws provided that at the relevant meetings of Council the medals should be dealt with ahead of any other item of non-routine business. In 1972 Council decided that it would be beneficial if there was a preliminary consideration of the awards by two subcommittees of Council, which would be asked to have a wide-ranging discussion of possible names and to bring a list of names forward for consideration by Council. Council members would still be free to propose names which had not been considered by the advisory panels, as they were named. It was hoped that the procedure would ensure that all worthy candidates were considered and that it would also save time at the November and December Council Meetings, which were traditionally dominated by discussions of the medals, because Council was having to deal with an increasing amount of other business. Two advisory panels composed entirely of Council members were set up, one to consider medals and Associates in astronomy and the George Darwin Lectureship and the other in geophysics and planetary Science to discuss medals, Associates and the Harold Jeffreys Lectureship. There were two members common to the panels to ensure a proper consideration of candidates whose fields of research and claims for recognition straddled the boundary between the panels.

Gradually during the decade Council changed its overall policy on medals, Associateships and lectureships. For a long time it was automatic for the George Darwin Lecture to be given by the astronomical Gold Medallist if he were not British. Sir James Jeans had expressed the wish that this would normally be the case when he endowed the lecture. At the same time outstanding foreign astronomers who were awarded medals were also extremely likely to be elected Associates as well. There was no such automatic connection between the Harold Jeffreys Lecture and the Gold Medallist, although he would sometimes be asked to give the lecture and there was no preference for a foreign speaker in that case. Council gradually moved towards the view that, while medals should be given for individual distinction in research, Associates should normally be involved in directorship of a successful institution, in important international collaboration or in some other activity which would not be a qualification for a medal, and that lecturers should be chosen for the quality and topicality of the lecture which they could be expected to give. There remained the expectation that some persons would be honoured in two or three capacities but that this would happen less frequently in the future than in the past. Although there were some outstandingly successful lecturers as a result of this new policy, it has to be admitted that there were also a few occasions when Council's judgement was at fault in the selection of a lecturer.

It was in 1971 that Council decided that it would no longer link the George Darwin Lecture with the award of the Gold Medal. In 1972 the Gold Medal was awarded to F. Zwicky but Margaret Burbidge was invited to give the George Darwin Lecture. She accepted the invitation but she subsequently withdrew this acceptance when she learnt that Zwicky had been awarded the Gold Medal but had not been invited to deliver the lecture. This problem did not arise in future years. In 1977, when the bye-laws of the Society were completely revised, Council took the opportunity to reintroduce the proposal that there should be Gold Medals specifically for geophysics and planetary science and for astronomy excluding planetary science and this time the proposal was accepted by the Annual General Meeting. Ever since the second medal had been introduced Council had worked with Standing Orders which almost ensured that result but it made things easier to have it specified in the bye-laws. The new bye-laws also opened Associateship to citizens of the Commonwealth and to United Kingdom citizens working abroad.

The Gold Medals

During the decade astronomical Gold Medals were awarded to the following people (noting that these were not technically astronomical until 1978):

1971　Sir Richard Woolley, for his contributions to observational and theoretical astrophysics, particularly in the field of stellar dynamics;

1972　F. Zwicky, for his many distinguished contributions to an understanding of the constituents of the Galaxy and the Universe;

1973　E.E. Salpeter, for his many distinguished contributions to theoretical astrophysics;

1974　L. Biermann, for his outstanding contributions to many branches of theoretical astrophysics;

1975　J.L. Greenstein, for his researches in astronomical spectroscopy and his inspiring leadership of the Department of Astrophysics at the California Institute of Technology;

1976　W.H. McCrea, for his brilliant pioneering contributions to several branches of theoretical astrophysics, including stellar atmospheres, cosmology, star formation and the physics of the interstellar medium;

1977　J.G. Bolton, for outstanding work in both radio and optical astronomy, particularly in connection with the initial discovery of discrete radio sources and subsequent optical identifications;

1978　L. Spitzer Jr, for his distinguished work in astronomy and, in particular, the study of the interstellar medium;

1979　C.G. Wynne, for a lifetime's work in geometrical optics as applied to astronomy, which has made possible many important types of observation;

1980　M. Schmidt, in recognition of outstanding contributions in astronomy, particularly investigations of the mass distribution of the Galaxy and of the nature of quasars.

In the same period geophysical and planetary science Gold Medals were awarded to:

1971 F. Press, in recognition of his distinguished personal researches in the field of seismology and the structure of the Earth's interior and of his outstanding leadership in earth and planetary sciences;

1972 H.I.S. Thirlaway, for his distinguished work in geophysics and for advancing seismological knowledge relating to earthquakes and explosions;

1973 F. Birch, for his work on the physical and chemical properties of the Earth's interior;

1974 K.E. Bullen, for his extensive and important contributions to theoretical geophysics, seismology and planetary physics;

1975 E.J. Öpik, for his pioneering researches in astronomy and particularly for his work on various aspects of the Solar System;

1976 J.A. Ratcliffe, for his outstanding contributions and worldwide influence in ionospheric physics and the theory of radio wave propagation;

1977 D.R. Bates, for his outstanding contributions to our understanding of atmospheric chemistry and dynamics and for his leadership in planetary and space science;

1978 J. van Allen, for his distinguished research in space science particularly in connection with the magnetosphere;

1979 L. Knopoff, for his many important and varied contributions to the physics of the Earth and especially to theoretical seismology;

1980 C.L. Pekeris, in recognition of outstanding contributions in geophysics particularly in the fields of seismology and oceanic tides.

It has already been mentioned to whom the first Chapman and Herschel Medals were awarded. Chapman Medals were also awarded to S-I. Akasofu in 1976 and E.N. Parker in 1979 and Herschel Medals to A.A. Penzias and R.W. Wilson in 1977 and G. de Vaucouleurs in 1980. Eddington Medals were awarded to D.G. King-Hele in 1971, P. Ledoux in 1972, S.W. Hawking and R. Penrose in 1975 and W.A. Fowler in 1978. The Society has one further award, the Jackson Gwilt Medal and Gift; the recipients were A.W.J. Cousins in 1971, G.E. Perry in 1974, P.A. Moore in 1977 and R.F. Griffin in 1980.

In 1975 there was concern about the rising price of gold and the Treasurer asked Council to consider whether it was still reasonable to spend the Society's general fund income on Gold Medals. Council decided that the expenditure was very worthwhile and not excessively high if a reasonably long historical perspective was considered. It was agreed that the Gold Medal should continue to be 9ct gold but that the Chapman, Eddington and Herschel Medals should be silver-gilt. Of course no change could have been made to the Gold Medal without a revision of bye-laws as it could not legitimately be called a gold medal if it was less than 9ct gold.

Council had one serious difficulty with regard to the George Darwin Lectureship. Academician V.L. Ginzburg of the Lebedev Institute in Moscow was selected to give the 1974 lecture. He accepted the invitation but was then unable to obtain permission to leave the USSR to come to a meeting of the Society to deliver it. Eventually it was decided that he would give the lecture in 1975 April or that, if he could still not obtain an exit visa, he would send the text to be read by someone else. In the event his lecture on 'Does astronomy need new physics?' was read by M.S. Longair.

Thoughout the decade three Associates were elected each year. There was an understanding in Council that at least one Associate would be in the field of geophysics and planetary science and, in fact, throughout the decade two astronomers and one geophysicist were elected yearly. The new Standing Orders which were adopted in the autumn of 1978 made the procedure clear and also laid down the guidelines for the choice of Associates which have been mentioned earlier. In future one Associate was to be an astronomer, one a geophysicist, with a free vote for the third. In 1978 the first United Kingdom astronomer working overseas to be elected an Associate was A.D. Thackeray, while at the same time M.K.V. Bappu was the first Associate who was a citizen of a Commonwealth country. Tragically, Thackeray died in a freak car accident after the decision had been taken by Council but before he learnt of it. There were two notable anniversaries relating to Associates: in 1972 May, C.G. Abbott celebrated his 100th birthday and G. Abetti was 90 in October of the same year. Each of them was sent a greetings telegram on behalf of the Officers and Council.

Leverhulme Visiting Fellowships

Through the initiative of Sir Bernard Lovell towards the end of his Presidency, Council was involved in another type of award for a period of five years. He obtained a grant of £5000 per annum for five years from the Leverhulme trustees to enable the Society to appoint two Leverhulme Visiting Fellows in Astronomy each year; the funds were subsequently increased to the extent necessary to take account of inflation to allow the full number of ten Fellows to be appointed at the envisaged seniority. The Fellows were to be from overseas, with some preference for less developed countries, and the Fellowships could be held at any university or research establishment in the United Kingdom. The first two Leverhulme Fellows, Dr T. Hirasawa from Japan and Dr R. Munro from Australia took up their Fellowships at Cambridge and

Manchester, respectively, in 1971. These Fellowships were very successful. There was a very strong field of applicants every year and all of the Fellows expressed pleasure at the opportunities that the Fellowships offered. It was very sad that Munro died just as his Fellowship was drawing to a close. Council was very grateful to the Leverhulme trustees and asked them to extend the grant for a further period. Although their response was initially encouraging, the Leverhulme trustees eventually decided that, in a time of accelerating inflation, they had too many other important calls on their available funds.

Meetings of the Society

The most important activity of the Society for many Fellows and Junior members, and particularly for those who lived within easy reach of London, was the meetings programme. Until just before the start of the decade there had been no significant change in the historical structure of the Society's meetings. The Ordinary Meetings of the Society were held from 16.30 to 18.30 on the second Friday of each month as was required by the bye-laws. The bye-laws authorised Council to organize additional meetings, which it did from time to time, the most important being the Geophysical Discussions, which were usually on Fridays between Ordinary Meetings and of which there were about four a year. There were occasionally extra Ordinary Meetings and meetings arranged for special purposes but the regular summer meeting away from London, which was a feature of the 1950s, no longer took place.

In 1969 the Secretary responsible for arranging meeting programmes, D. McNally, recognized that the increasing costs of travel and rising pressure of other commitments made it difficult for members to justify coming to London from a distance for only two hours' science, particularly as there were so many other competing scientific meetings which they could attend during the year. What had been an ideal arrangement when the Society had an extremely large amateur element living in or near to London was less attractive to the increasing number of young professionals. Following McNally's suggestion, Council introduced occasional early Friday afternoon discussion meetings devoted to specific topics, while the Ordinary Meetings continued to contain a wide variety of subjects. Soon there were discussion meetings on every meeting day except the day of the Annual General Meeting.

These discussion meetings were so successful that a later Secretary, J.R. Shakeshaft, suggested that starting in 1974 December the discussion should start at 11.30 with a break for a sandwich lunch. In putting forward this proposal he said that participants felt that there would be increased opportunity for discussion at a longer meeting including a lunch break and that it would also help them obtain travelling expenses.

The extended discussion meetings were immediately adopted and, looking slightly beyond the end of the decade, they were so popular that in the early 1980s there was a tendency to have parallel astronomical and geophysical meetings. At the same time the starting hour edged forward to 11.00 and then 10.30. To make the meetings even more attractive, Council was prepared to pay the expenses of a participant from overseas, if the organizers felt that this would enhance the value of the meeting. As a result the meetings of the Society had an increasingly international flavour.

Out-of-town meetings

One of the consequences of the substantial increases in the Annual Contributions of Fellows and of corresponding increases in Fellows' prices for the two main journals was that there was serious questioning of the value of membership of the Society for those members who lived too far from London to be able to attend the Society's meetings regularly. This led to lengthy discussions in Council in 1975 concerning the role of the Society and one outcome of these discussions was a decision to reintroduce out-of-town meetings. This was influenced by the replies to a questionnaire sent to all members asking them for their comments on specific activities of the Society. 1100 members replied and one strong request was for more meetings outside London. The replies also indicated that most members were pleased to receive both *Quarterly Journal* and *The Observatory* and to pay for them in their Annual Contribution. Council felt that the vast increase in the number of astronomical symposia and colloquia in recent years made an additional summer meeting less attractive than in the past. Instead it was decided to try the experiment of holding the April Ordinary Meeting away from London and to associate discussion meetings and possibly other activities with it.

The Society met in Manchester in 1976 and the meeting was so successful that it was followed by further meetings in Edinburgh in 1977, York in 1978, Durham in 1979 and Cardiff in 1980. Associated with each meeting was a lecture open to the public which was, on occasion, one of the Society's named lectures. In arranging these meetings, Council relied on local organizers to do most of the work with some financial support from the Society. There proved to be no difficulty in finding local organizers who were keen to have a meeting and the meetings were well attended and very enjoyable for the participants. One problem in arranging out-of-town meetings was difficulty in ensuring that the April Ordinary Meeting should be held on the second Friday, or, if that were Good Friday, on the Wednesday preceding it. Eventually the bye-laws were changed to allow Council to arrange the

Ordinary Meeting on a different day provided six months notice was given.

Geophysical Meetings

A recurrent item of Council business throughout the decade involved the relationship between astronomy and geophysics within the Society. Some of the organizational aspects will be mentioned below; here some developments concerning meetings will be dealt with. Before the period under review, most geophysical activities were concentrated in the Geophysical Discussions held on Fridays other than Ordinary Meeting days and in occasional special and extra meetings. The Ordinary Meetings were devoted almost entirely to astronomy with exceptions, for example, when the Harold Jeffreys Lecture was given. The introduction of discussion meetings on Ordinary Meeting days led to the attempt to hold Geophysical Discussions on those days and to make the Ordinary Meetings of interest both to astronomers and to geophysicists so as to bring the Society together. This could not be described as an outstanding success, although shortly after the end of the decade the attempt was still being made with parallel astronomical and geophysical discussions being held on the same day. One problem was the tendency for large numbers of people to leave after the first talk at the Ordinary Meeting, if that was the only talk of particular interest to them. This was part of a wider tendency for attendance at the Ordinary Meeting to be regarded as an optional extra. Although there was still a sizeable group of Fellows who came for the Ordinary Meetings alone, many others came primarily for the discussion meeting and stayed for part of the Ordinary Meeting if it was in their own or a closely related field.

One group of geophysicists primarily associated with the Earth's ionosphere and related problems developed their own series of meetings. The meetings on the Magnetosphere, Ionosphere and Solar–Terrestrial Relations (MIST for short) were usually held on days other than Ordinary Meeting days and very often in association with the Institute of Physics and with some financial support from the Society. The scientists who regularly attended these meetings formed a coherent group and they seemed well satisfied with their two or three meetings a year. Although they were such an active group inside the Society, they did not as a rule submit their researches to *Geophysical Journal* which continued to be primarily a journal for solid earth geophysics.

An important development in geophysical meetings in which the Society played a role was the decision to hold an annual United Kingdom Geophysical Assembly (UKGA). The Society sponsored the first meeting held at Edinburgh in 1977; it underwrote the financial arrangement to some extent at no eventual cost to the Society and it also published the abstracts of the papers in an issue of *Geophysical Journal*, which appeared just before the meeting. The first UKGA was extremely successful and further meetings were held at Liverpool in 1978, Southampton in 1979 and Birmingham in 1980. The assemblies were arranged without there being any permanent bureaucracy associated with them, which seemed particularly advantageous. All of these assemblies were held in April like the out-of-town RAS meetings. This raised some minor problems when the organizers of the out-of-town meetings also wished to include some geophysics so that the meeting would cater for the whole Society.

A further important development was in collaboration with the Geological Society in fields of common interest. In 1976 Council issued a standing invitation to Fellows of the Geological Society to attend geophysical meetings organized by the Society and it also agreed to a reciprocal arrangement for the advertisement of geophysical meetings organized by either society. In 1978 a further development occurred when it was arranged that members could receive the *Geological Society Newsletter* in place of *The Observatory* if they so wished. A number decided to do this, although it was noticeable that many members, who would be primarily classified as geophysicists, continued to receive *The Observatory*. The most important development was the formation of the Joint Association for Geophysics (JAG), membership of which was open to members of either society. JAG had a committee drawn equally from members of the two societies and the idea was that it would co-ordinate geophysical meetings. A by-product of the formation of JAG was the agreement that members of either society could buy journals of the other society at a special price. One worry about this arrangement was that it might cause the more geologically minded Fellows to resign from the RAS once they could obtain *Geophysical Journal* through the Geological Society. Initially there were a few losses but not enough to cause serious concern. JAG started its activities only in 1979 and it was too early by the end of the decade to estimate what its influence would be likely to be.

Special Meetings

Special note should be taken of some of the Society's meetings. 1971 was the quatercentenary of the birth of Kepler and it was decided to devote the December Ordinary Meeting to a celebration of the occasion. Both the George Darwin Lecture and the Harold Jeffreys Lecture were delivered on that day. The George Darwin Lecture was 'Johann Kepler and the new astronomy' by Professor Owen Gingerich and the Harold

Jeffreys Lecture was 'Heavenly harmony and earthly harmonics' by Mr D.G. King-Hele. This meeting was held in association with the Royal Society and the British Society for the History of Science. In the following year a discussion meeting on supernovae celebrated the four hundredth anniversary of Tycho's supernova of 1572 and the principal speaker was the Gold Medallist of the year, Professor F. Zwicky.

The decade was one of great development in the provision of facilities in both ground-based and space astronomy, although by the end of the period a change in the financial climate was beginning to become apparent. The Society took the initiative in proposing to the Science Research Council (SRC) that it might be appropriate to hold occasional meetings at which representatives of the SRC could discuss its plans with the astronomical community. Two such meetings were held in 1974. The first, on March 1st, was a general description and discussion of the plans of the SRC and a second meeting, on June 21st, was devoted to plans for a Northern Hemisphere Observatory.

At the meeting on March 1st the principal speakers were Sir Eric Eastwood, Chairman of the Astronomy, Space and Radio (ASR) Board, Dr H.H. Atkinson, Director of the Astronomy Space and Radio Division, Dr D.W. Sciama, who spoke about the science which the ASR Board hoped to encourage, and Professors D.W.N. Stibbs and H. Elliot, Chairmen of the Astronomy Policy and Grants Committee and the Space Policy and Grants Committee, respectively. The Chairman and Secretary of the SRC, Professor S.F. Edwards and Mr R. St J. Walker, were also present. After the formal presentations there was a vigorous discussion about both the organization of the SRC's support for astronomy and the science which might be undertaken with the facilities to be provided. The report of the discussion occupied ten pages of the 1974 December issue of *The Observatory*. The speakers at the meeting on June 21st included Professors D.W.N. Stibbs, R.O. Redman and J. Ring and Mr B. McInnes, who were all members of the Northern Hemisphere Observatory Planning Committee.

The Ordinary Meeting of 1977 October 14 was dedicated to the memory of Sir James Jeans, the centenary of whose birth fell in September. A talk about Jeans was given by Professor W.H. McCrea and there were three talks about recent developments in subjects in which Jeans was a pioneer. The Society was happy that Lady Jeans and other members of the family could be present. In 1979 March a similar commemorative meeting was held when the Ordinary Meeting celebrated the centenary of the birth of Einstein. There were seven contributions on his work and on his association with the RAS and an exhibition of related archival material was mounted by the librarian. In 1980 a successful innovation was a Junior Members' Day at the May meeting where 16 junior members gave presentations of their research.

The general standard of these talks was very high and Council expressed an intention to hold similar meetings in the future. The Society was associated with a special geophysical meeting on 1974 June 28 when it co-sponsored the retirement meeting of Sir Edward Bullard at Cambridge.

Structure of Ordinary Meetings

There should be a few remarks about the general organization of Ordinary Meetings. When the decade started, the meeting was still very much in its traditional form. Formal business at the beginning of the meeting included the election of Fellows and Junior Members at which Fellows were invited to demand a ballot, if they so wished. On several occasions the officers feared the worst when some Fellow inadvertently raised his hand but fortunately no ballots were demanded. What would have happened if a ballot had been demanded at a January meeting when more than a hundred Fellows were to be elected is difficult to imagine. Other formal business included the reading of papers communicated, presents received and names of members proposed. All these items had been considerably abbreviated. When the bye-laws were revised in 1977, the need for all this business was removed, Council being given the power to elect Fellows; their names continued to be suspended in the Society's premises for a month to give Fellows an opportunity to object to any particular proposal. The only remaining item of formal business was the admission of Fellows and Associates.

There was also a gradual change in the nature of the Ordinary Meeting. Traditionally the meeting was devoted to the reading of papers received by the Society and it was expected that these papers would subsequently be published in the Society's journals. Occasionally a Secretary would invite an author to deliver a paper which subsequently had such trouble with referees that it never appeared, but that was rare. In addition to the routine papers there were Presidential Addresses, addresses on the award of medals and George Darwin and Harold Jeffreys Lectures. All of these would also subsequently be published. Only occasionally was there a talk at an Ordinary Meeting, which was specially arranged because it was on a topical subject. Such a talk would probably not subsequently be published. For a long time there had been many more papers communicated to the Society than could possibly be read at the Ordinary Meetings, although in previous decades authors were formally asked whether they would like to present their papers. Although papers did continue to be read at Ordinary Meetings in the 1970s, there was an increase in the proportion of time devoted to special talks. A further change was that the Presidential Addresses on the award of the medals ceased to be given at the February meeting. Instead it was decided that the address should be given when the medal was presented so that the recipient would always hear it. Finally a major break with custom took place in the autumn of 1979 when it was decided that in the future Ordinary Meetings would start at 16.00 and end at 18.00.

Venue of Meetings

Most of the Ordinary Meetings during the decade were held in the Scientific Societies Lecture Theatre in Savile Row. Although there was regret that the Society could no longer meet in its own premises and that the gathering and tea before Ordinary Meetings could no longer be held in the library, there was general satisfaction with the facilities at Savile Row, where the meeting room was large enough for a greatly increased attendance. Satisfaction with these facilities was reduced when the Department of the Environment, which was responsible for the building which housed other offices as well as the lecture theatre, decided that security considerations would no longer permit astronomers and their guests to enter through the front of the building. Instead a new entrance was made in a side passage. This had two disadvantages. The entrance to the lecture room was now at the front which made late arrivals and early departures more obvious and inconvenient. Outside, the door was next to the building's dustbins. This led to complaints in Council and at Annual General Meetings but attempts to persuade the DOE to re-open the original access or to provide an improved one were unsuccessful. Despite the Society's earlier investment in the lecture theatre, Council tried to find a suitable alternative meeting room. In 1979 a tentative decision was taken to meet more frequently at the Royal Society at Carlton House Terrace but a complete move could not be made because they would not accept a regular booking.

The Officers

The Presidents

The decade started with the end of a memorable Presidency. Sir Bernard Lovell was President from 1969 to 1971 and 1970 had seen not only the Sesquicentenary of the foundation of the Society but also the holding of the XIVth General Assembly of the International Astronomical Union at Brighton. There were five other Presidents during the decade, Sir Fred Hoyle (he was knighted during the first year of his Presidency), Professor D.E. Blackwell, Professor F. Graham Smith, Professor A.H. Cook and Professor M.J. Seaton. All of these Presidents in their different ways served the Society very well and between them they gave a number of memorable Presidential Addresses. The full titles of and references to the published versions of the addresses are given in Appendix 1. The Report of Council to the Annual General Meeting in 1974 noted that on 1973 February 9 the Scientific Society's Lecture Theatre was packed with more than 300 members and guests for Sir Fred Hoyle's Presidential Address on 'The History of the Universe'. One thing that became clear in the 1970s was that the President was quite definitely a working officer. It did not seem possible to elect someone as President who would be simply a figurehead. Professor Seaton had a particularly difficult period of office because of the successive deaths of an Executive Secretary and a Secretary but he impressed everyone by being the perfect President for such a difficult time.

Lovell is a radio astronomer, one of the pioneers of the subject, and is very well known to the general public as well as to other scientists through the success of the 250 ft radio telescope at Jodrell Bank, which was still the largest single-dish radio telescope in the world at the start of the decade. Although his telescopes were probing the ultimate depths of the Universe, he also had a personal research interest in radio emission from red dwarf stars, a topic of one of his Presidential Addresses.

Hoyle can only be described as imaginative, stimulating and controversial, with research interests covering theoretical problems in most of astronomy and also touching on geophysics. He surprised the astronomical community by resigning his Cambridge chair in the middle of his Presidency and by the end of the decade he was stirring up new controversy through his ideas on the origin of life in space.

Blackwell started work in optical astronomy but when he was President his work was primarily in laboratory astrophysics. He was carrying out painstaking measurements of atomic constants, f-values, which are crucial to an understanding of such things as the chemical composition of stars.

Graham Smith is a radio astronomer but he was involved in the organization of optical astronomy during his Presidency. He was at first Director-Designate of the Royal Greenwich Observatory and he succeeded Dr A. Hunter as Director in 1976 January. His main concern was with the arrangements for setting up the new international observatory at La Palma in the Canary Islands and with procuring the United Kingdom telescopes for the observatory.

Cook is primarily a geophysicist and was one of the founders of the

Geophysical Journal, but he has very much wider research interests, including molecular astronomy, and his book on celestial masers was published in the year that he became President.

Seaton's principal research work is in theoretical atomic physics and in the use of these calculated atomic parameters in understanding the observed properties of gaseous nebulae.

The Presidency of Graham Smith included the celebrations of the tercentenary of the Royal Greenwich Observatory. Although the Society is much younger than the Observatory, their relations have been extremely close. In 1830 the Board of Visitors of the Observatory was reconstituted to include the President and five Fellows of the Society, with the President to act as Chairman in the absence of the President of the Royal Society and this arrangement continued until the Board ceased to exist when the Science Research Council took over the Observatory in 1965. In addition five successive Astronomers Royal, Airy, Christie, Dyson, Spencer Jones and Woolley, had been Presidents of the Society. Dr A. Hunter, as Director of the Royal Greenwich Observatory, invited Fellows and their guests to visit it on a special open day on 1975 June 24 and 280 Fellows and guests were present. One of them was Professor J.L. Greenstein, the 1974 Gold Medallist, and the President took the opportunity to present the medal and to invite Greenstein to lecture on 'The Stars of Low Luminosity'. It was particularly appropriate that the President was himself Director-Designate of the Observatory and that Hunter, who had been awarded a CBE in the Birthday Honours List, was Treasurer of the Society. The tercentenary was also celebrated by a joint Royal Society/Royal Astronomical Society reception for participants in a historical symposium at Greenwich, which took place at the Royal Society on 1975 July 17.

In 1980 February Seaton presided at the 160th Anniversary Dinner of the Society at University College, London. Professor W.H. McCrea proposed a toast to astronomy and geophysics and the President of the Geological Society responded. This was a very successful and enjoyable occasion but the contrast with the Sesquicentenary dinner held at the Dorchester Hotel said something about the inflation of the decade.

While we are discussing social events, it is an appropriate point to record that starting in 1975 occasional sherry parties were held in Burlington House after an Ordinary Meeting. The parties were self-financing and proved very popular, the Fellows' room being packed to bursting point. This enabled Fellows and guests to have further informal discussions and they also ensured that significant numbers of members visited Burlington House now that the scientific meetings were held elsewhere. In time the sherry parties were held every month and in 1979 November it was agreed that they would in future be publicized on the meeting notices.

Treasurers and Secretaries

At the beginning of 1971, A. Hunter was coming to the end of his first four years as Treasurer of the Society. He continued as Treasurer until the Annual General Meeting of 1976. His experience of the first four years could not have prepared him for the rapid inflation of the later years during which he had the unenviable task of persuading the Fellowship to agree to an increase of the Annual Contribution by a factor of 3 between 1973 and 1976. It is however clear that this secured the financial equilibrium of the Society. Hunter was succeeded in 1976 by W.H. McCrea, who served until 1979. A particular feature of McCrea's Treasurership was his success in obtaining financial support from outside bodies, particularly for the library and for its care of the Society's oldest books. R.J. Tayler became Treasurer in 1979.

The decade started with three Secretaries, T.F. Gaskell, F. Graham Smith and D. McNally, who had already been in office for some years. R.J. Tayler succeeded Graham Smith in 1971 and J.R. Shakeshaft and J.A. Hudson succeeded McNally and Gaskell in 1972 and 1973 respectively. There followed a period of five years with no change in Secretary. J.A. Jacobs and R.D. Davies followed Hudson and Shakeshaft in 1978 and V.P. Myerscough took over from Tayler in 1979. Unfortunately, Valerie Myerscough, who had earlier served the Society very well as Convener of the House Committee, as will be mentioned later in this chapter, died in the autumn of 1980 and she was succeeded by C. Jordan right at the very end of the decade.

One officer and six former officers died during the decade. The early deaths of the Secretary, V.P. Myerscough, and the former Secretary, A.D. Thackeray, have already been recorded. Although Thackeray had been Secretary for two brief periods (1941–43, 1947–48), he had spent the major part of his astronomical career in South Africa as Director of the Radcliffe Observatory and that prevented him from playing what would surely have been a major role in the activities of the Society. Five former Presidents also died, R.O. Redman, W.M. Smart and W.H. Steavenson in 1975, H. Dingle in 1978 and H.H. Plaskett in 1980. With the exception of Redman they had reached a good old age and their deaths were not unexpected, but Redman died tragically quickly from leukaemia when he had been retired for less than three years and when he was still active in plans for the new large United Kingdom optical telescopes. Unfortunately Dingle did not live to see his contribution to this volume published.

Council Business

The volume of Council business grew during the decade in spite of a reduction in some of the business which had occupied much of Council time earlier; for example, the institution of the Advisory Panels on Medals and Awards meant that much of the preliminary discussion of possible candidates took place in the Advisory Panel meetings rather than in Council. In addition, once the Editorial Boards had been authorised to reject papers, there was little discussion in Council of the details of papers submitted to the Society. Despite this, whereas in 1971 Council held eight meetings with an average duration of 1 hour 50 minutes, by 1980 there were nine meetings with an average length of more than 2½ hours. The extra meeting was a July meeting because it was now necessary to spend time discussing the Society's budget and the prices of publications every year and July was the last month in which this could reasonably be done.

The time of Council meetings also changed twice. At the start of the decade it was at 11.00, having only quite recently moved from 14.00 as a result of the introduction of afternoon discussion meetings. One consequence of the discussion meetings and the earlier time of Council was that it was no longer possible for the Society's committees to meet in the morning of the Ordinary Meeting day. The officers met at 10.00 and this allowed officers and Council members to attend both the discussion meeting and the Ordinary Meeting in the afternoon. Following the introduction of morning discussion meetings in 1974, the time of the officers' meeting was brought forward to 09.15 and Council met at 10.00 with the hope that it would be possible for Council members to attend the discussion meeting at 11.30. In particular it was expected that the most important Council business would finish before 11.30, so that any Council member closely involved in the organization of discussion meetings would be able to leave Council early if necessary. This hope was not really fulfilled, particularly as discussion meeting organizers were continually pressing for their meetings to start earlier. In 1979 October Council recognized that it was virtually impossible for Council members to attend both Council and a morning discussion meeting and changed its meeting hour to 10.30. Officers continued to meet at 09.15, partly because they really did seem to have an increasing amount of business to deal with.

As has been mentioned earlier more than once a major concern of Council was the relationship between astronomy and geophysics in the Society. There was clearly a tendency for some geophysicists to feel that they were being treated as second class members of the Society, while at the same time the *Geophysical Journal* continued to earn a very large fraction of the Society's financial surplus. One problem was that of ensuring that Council contained a large enough representation of geophysicists to look after their general concerns as well as to give adequate advice on the selection of geophysical medallists, lecturers and Associates. Another problem was that there was not really one geophysics as we have already seen when discussing the MIST meetings and the Joint Association for Geophysics.

The problem of Council membership in fact came to a head in the last year of the previous decade, 1970. Council proposed a nearly complete list of nominations, as was usual, for the new Council, containing sufficient geophysicists for a balanced Council. There were, however, several additional nominations of astronomers and two of them were elected in place of two of Council's own geophysical nominees. This was particularly embarrassing as earlier in the same year only one Gold Medal had been awarded, to an astronomer. This was the first and so far only occasion since the second Gold Medal had been instituted that no medal was awarded in geophysics. After this election experience Council ceased to propose so many nominees for the new Council. In its list it included what it regarded as a satisfactory mix of astronomers and geophysicists with the hope, which was fulfilled, that the genuine election would be for the remaining places on Council. In the reform of the bye-laws in 1977, provision was made that at least three of the 12 ordinary members of Council should be geophysicists and at the same time the aims of the Society were specifically stated to be the furtherance of astronomy and geophysics.

The new rules relating to the composition of the Council complicated the task of the scrutineers of the ballot and in 1980 a transcription error led to the Annual General Meeting being told that M.J. Rycroft had been elected to Council, when the correct name should have been A.W. Wolfendale. The error was discovered immediately after the meeting and it is a pleasure to report that Rycroft was elected to Council in the following year.

The revision of the bye-laws has been mentioned several times already and this was a matter which took up a large amount of Council time in the middle part of the decade. When it became clear in 1975 that there were several reason why bye-law revision was desirable, Council decided to take the opportunity to look at all of the bye-laws with the view of simplifying them and modernizing their wording as far as possible in the hope that only minimal changes would subsequently be needed for some time. A Bye-laws Committee was set up at the 1975 May Council meeting with the initial hope that it would report in time for the bye-laws to be proposed at the 1976 Annual General Meeting. In fact it soon became clear that this time-scale was too short. Although a semi-final draft of the new bye-laws was considered at the 1976 April meeting, the final draft was only just available in time to submit the proposal to the 1977 Annual General Meeting. Here, the revision was accepted with a few last-minute amendments, which were acceptable to Council. One item which caused much heart searching was whether to propose that in future the Annual Contribution would be determined by Council without reference to the Fellowship, as was the case in some other learned societies. This proposal was contained in one draft but eventually it was replaced by a requirement that Council must propose the Annual Contribution to the General Meeting.

In discussing publications, meetings and medals we have already dealt with many of the most important items of Council business. Some other items of a domestic nature are as follows. In 1973 May Council considered complaints that a Fellow had been spoiling the enjoyment of meetings by other members by numerous noisy interjections. The officers even went so far as to look into the procedures for expulsion of a Fellow should that prove necessary. Fortunately the Fellow concerned ceased to attend meetings and was shortly afterwards in default with the Annual Contribution and ceased to be a Fellow. In 1974 November an important decision was taken which related to travelling expenses of Council members. For a long time members of Council and the Society's committees were entitled to claim a first class rail fare for meetings which were not on Ordinary Meeting days, when no expenses were payable. It was assumed that all active members of the Society would in any case be attending the Ordinary Meetings. In 1974 November it was agreed instead that only second class rail fares would in future be paid but that Council members could claim the expenses for attending Council less £3.50. The latter figure was chosen to exclude Councillors from Oxford, Cambridge, Sussex and London, whom it was thought could still reasonably be able to afford to attend regularly. For the subsequent two years, the base figure was revised each year as rail fares increased until in 1977 it was agreed that in future all Council members could claim their full travelling expenses.

In 1975 January it was agreed that the Society would produce a booklet on postgraduate opportunites in astronomy and geophysics on a self-financing basis. Departments which admitted postgraduate students were invited to submit material in camera-ready form for one or more pages in the booklet. They were also asked to pay a page charge in return for which they would have copies of the booklet for their own use. This booklet was widely welcomed and by the end of the decade it was determined that it would be updated every two years.

In 1977 July a decision was made to transfer the Society's copy of the Palomar sky survey on permanent loan to University College, London. The sky survey was a gift of Mr Jack Miller and his generosity was very much appreciated by the Society. By 1977 the European Southern Observatory and UK Schmidt southern sky surveys were being produced and it was clear that the Society could not really expect to acquire and house them. As University College, London was going to receive copies of these two southern surveys and as it contained a very large concentration of astronomers, it seemed appropriate that the Society's Palomar survey should be located close to the southern surveys in a place where it would be much more regularly used than it had been when on the Society's premises. A condition of this transfer was that members of the Society would be allowed access to both the northern and the southern surveys.

In 1978 two financial matters, which have not already been discussed, were settled. The first and potentially very important one was whether the Society would be liable to pay Value Added Tax, particularly in respect of the Annual Contributions of Members. In 1978 March Council was pleased to hear that the Executive Secretary had obtained a ruling that the Society could claim exemption from VAT on subscriptions. Later in the year the Executive Secretary also made an arrangement with the Inland Revenue whereby Fellows who covenanted their annual contributions to the benefit of the Society could sign a variable covenant which would not have to be changed each time the Annual Contribution changed. Following this, an appeal for new covenants was very successful.

The Loyal Address

In 1976 December, Council considered what it should do to recognize the impending Silver Jubilee of the Society's Patron, Her Majesty the Queen. It was decided that she should receive a loyal address from the Society and Dr D.H. Sadler was asked to prepare a draft for consideration by Council. The first draft was available at the February meeting of Council, the final text was settled in March and the address was produced for presentation in 1977 May. At the July Council meeting a reply to the loyal address was received. The text of the address was as follows:-

> To Her Most Excellent and Gracious Majesty Queen Elizabeth
> May it please your Majesty
>   We the President, Council and Members of the Royal Astronomical Society tender our respectful congratulations on the happy occasion of the completion of the twenty-fifth year of Your Majesty's reign.
>   Twenty-five years ago Your Majesty was graciously pleased to become our Patron, and we desire to express our deep appreciation of the honour so conferred on the Society. The period since then has been the most fruitful quarter-century for astronomy in the history of our science, in it our knowledge and understanding of the Earth, the Solar System and the Universe have increased immeasurably. We appreciate with gratitude the lively interest that Your Majesty and other members of the Royal Family have shown in these advances in astronomy and its related sciences and particularly in the inauguration of new and powerful observational facilities

in Great Britain and Overseas. We trust that the Society may long enjoy the privilege of Your Majesty's patronage.

With homage and esteem we pray that in peace and happiness Your Majesty's reign may long continue.

Presented 1977 May 13
ALAN COOK
President
ROGER J. TAYLER
Secretary

A photograph of the address and the texts of both the address and reply were published in *Quarterly Journal* of 1977 December.

National and International Affairs

From time to time Council was asked to express an opinion on matters which went beyond the concern of the Society. On such occasions Council had to balance the desirability that the voice of the Society should be heard against the necessity of reflecting the opinions and concerns of the whole membership of the Society; for example, it would not be appropriate for Council to urge the claims of ground-based astronomers against those of space astronomers, or of astronomers against geophysicists. In addition Council had to avoid making political statements which might offend many members. In 1977 Council was, however, unanimous in asking the President (A.H. Cook) to write to the Chairmen of the Science Research Council (SRC) and of its Astronomy Space and Radio Board to express the high importance placed by Council on the acquisition of a 4.2 m blank from Owens Illinois for the proposed large telescope on La Palma, Canary Islands. The purpose of the letter was to help the SRC itself to persuade the Department of Education and Science to allow it to put down a deposit on the mirror blank before the establishment of the new observatory had been formally agreed. The blank was subsequently acquired.

When the decade started, Council nominated assessors to the SRC Astronomy Space and Radio Board and to the two committees immediately under it, the Astronomy Policy and Grants Committee (APGC) and the Space Policy and Grants Committee (SPGC). It also nominated three members of the Royal Greenwich Observatory Committee and one member of the Royal Observatory Edinburgh Committee. In 1974 the SRC committee structure was changed. The APGC and SPGC were replaced by three committees, Astronomy I, Astronomy II and Solar System, on each of which the Society had an assessor. Subsequently, in 1976, the two observatory committees were replaced by renamed Establishment Committees with revised terms of reference and an

Appleton Laboratory Establishment Committee was added. The Society now had an assessor on each. The role of the assessors was to keep Council and through it the Society and the astronomical community aware of SRC plans for astronomy and the part of geophysics for which it was responsible. This became more important towards the end of the decade when finances became tight. In 1980 Council tried to persuade the Natural Environment Research Council (NERC) to allow it to appoint an assessor to an appropriate NERC committee but in this it was unsuccessful. Council also nominated members of Royal Society National Committees for Astronomy, Geodesy and Geophysics, Geology, Oceanic Research and Space Research and to the Optical and Quantum Electronics Sub-Committee of the Physics Committee. In 1978, after the SRC moved its headquarters from London to Swindon, it was encouraged to use the Society's Council Room for meetings of some of its smaller committees, an appropriate rental being charged.

In 1979 January Council was asked to intervene with the SRC by many members of the staff of the Appleton Laboratory who were objecting to its proposed amalgamation with the Rutherford Laboratory and to their transfer from Slough to Chilton. Council agreed to consult its assessor on the Appleton Laboratory Establishment Committee, Professor K.A. Pounds, and he advised Council at its next meeting that there was no point in pursuing the matter further as SRC Council had already taken its decision.

In 1980 April a member of Council, Dr S.J. Bell Burnell, brought to the attention of Council the problems arising from an SRC rule which usually restricted to three years the employment of post-doctoral assistants on research grants. This not only raised problems for individuals whose research careers might be terminated but it also sometimes caused difficulties for projects. As a result Council produced some preliminary views which were presented to the ASR Board on 1980 November 25. Council's recommendations were that: while the normal period of a PDRA appointment should be for three years, appointments should be extended to six years in cases of excellence and timeliness in research and that there should be no absolute limit to length of appointment; long-term appointments should be available for scientists whose main contribution is in providing a service to research groups; the SRC should encourage universities to appoint more astronomers to permanent posts. No significant increase was sought in the number of man years of PDRA support but rather a more flexible use of the funds available. At its December meeting Council set up a panel to consider employment prospects in astronomy; the panel's conclusions are a subject for a future volume of the *History*.

On 1980 February 8 Council responded to a request by another Fellow, Dr M.K. Wallis, that it should intercede with the USSR Academy of Sciences on behalf of Academician A.D. Sakharov, whose movements had recently been restricted. On this occasion Council decided that it could reasonably intervene in a political matter and the following telegram was sent to the Academy:

> The Council of the Royal Astronomical Society notes the importance of the scientific work of Academician Sakharov and the relevance of this work to current research in astronomy.
>
> The Council considers the restrictions recently placed on Academician Sakharov to be a violation of human rights. These restrictions threaten relations between Soviet and British astronomers and geophysicists and endanger the continuation of scientific exchanges and the holding of international conferences in the USSR.
>
> The Council of the Royal Astronomical Society appeals to the Academy of Sciences of the USSR to use its influence to secure the release from exile of Academician Sakharov.

The text of the telegram was read out at the Ordinary Meeting that afternoon and it was subsequently broadcast in Russian by the BBC East European Service.

In 1971 Council had several discussions about proposed collaborations in European astronomy. There was a proposal that a European Astronomical Society should be formed and details of the proposal were communicated to the Society by Professor G. Contopoulos. Council was not enthusiastic about the proposed society as it felt that there was no need for a new tier between national societies and the International Astronomical Union (IAU). It was suggested that the IAU should itself promote regional activities and that was later arranged in years when there was not an IAU General Assembly. Also in 1971 there was discussion of the astronomy section of the European Physical Society (EPS) and whether arrangements could be made for the Society's members to join it at a special subscription rate. Investigation showed that EPS corporate membership was too expensive because it involved a capitation charge based on the full membership of any joining society. In the RAS case this membership includes many amateurs, geophysicists and overseas members, for whom the EPS astronomy section would be largely or wholly irrelevant, so no action was taken.

Finally on 1971 September 13 there was a meeting at Bonn of representatives of the Society, of the Astronomische Gesellschaft and of the German Federal Republic National Committee for Astronomy with the General Secretary and the Assistant General Secretary of the IAU to discuss European co-operation. It was suggested that the Society and Astronomische Gesellschaft should jointly sponsor occasional specialist meetings and that there might be an occasional exchange of visits by distinguished astronomers from the two countries. In fact nothing tangible resulted from this meeting, although the Society did encourage an increased participation of European scientists in its own meetings.

There was a small international interaction in 1974, when the Society was asked to nominate a young astronomer to be present at the inauguration of the Anglo-Australian Telescope and Council nominated Dr M.A. Dopita. There were two cases in which Council was able to help institutions overseas, which had suffered losses as a result of fires. In 1979 March it was agreed to supply some replacement copies of the *Geophysical Supplement to Monthly Notices* and of the *Geophysical Journal* to the University of Lisbon, while in December of the same year it was agreed to provide some replacement copies of *Monthly Notices* to Dunsink Observatory.

In the late winter and early spring of 1971 the activities of the Society and indeed of the whole United Kingdom were seriously impeded by a postal strike, which lasted about eleven weeks. Publication of the journals and the refereeing of papers was seriously delayed and special arrangements had to be made for the delivery of such items as Council meeting papers. As the period included the Annual General Meeting there were worries about whether there would be a representative vote for officers and Council. Council was particularly indebted to Professor P.A. Wayman of Dunsink Observatory who received the ballot forms from overseas Fellows at his observatory and then delivered them personally to Burlington House.

Also in 1971 two international meetings on planetary science were held in the United Kingdom, an IAU Symposium on The Moon on March 22–26 and an International Colloquium on The Earth and Planetary Physics on March 30–April 2. Professor S.K. Runcorn suggested and Council agreed that a reception for overseas participants should be held at the Society's premises on March 29. On this occasion the exhibition which had been mounted to celebrate the Sesquicentenary in 1970 was reassembled.

Further matters which were discussed by Council will come up in later sections of this chapter.

The Society's Staff

During the decade the Society was very well served by three Executive Secretaries, to use the title which replaced that of General Secretary in the bye-law revision of 1977: Edward Rubidge who held the office with a succession of titles from 1956 until his retirement in 1977 December; Alan Hume OBE from 1978 January until his untimely death in 1979 December; and Tony Steff-Langston MBE from 1980 May. In the period from 1980 January to May, Mrs Enid Lake, the Librarian, was also a very efficient Acting General Secretary.

In the early part of the decade staffing was extremely difficult. There were three reliable and long-serving members of staff, Rubidge, Dr R.E.W. Maddison (Librarian) and Mrs M.M. Bassett (Assistant Secretary), together with the Caretakers, Mr and Mrs Hudson. Unfortunately Mrs Bassett was in very poor health and she died in 1973 December. A reading of Council minutes shows that the additional members of staff tended to come and go almost too rapidly to be recorded there. As an example, between 1971 May and 1972 April no fewer than four members of staff were appointed and later resigned. I first became a Secretary of the Society in 1971 March and it is my recollection that the staff in the Society's general office was rarely the same from one meeting to the next. As a result of Mrs Bassett's ill-health and the unreliability of junior staff, there were times when Rubidge appeared to be running the Society almost single-handedly. Frequently Maddison had to do work which was not really the concern of the Librarian. At the beginning of each of 1973 and 1974 Council made an additional payment to Rubidge in token of the very long hours he had been working and such a payment was also made to Maddison in 1973. Naturally officers and Council were concerned about whether the Society's salary scales were adequate to attract suitable staff; as they were tied to Civil Service scales it was felt they should have been satisfactory. In the middle of the decade a transformation occurred and almost complete stability in the Society's staff was achieved by the time that Edward Rubidge retired and he was able to hand over a very reliable staff to Alan Hume.

The services which Edward Rubidge gave to the Society cannot be exaggerated. He worked in effect as Executive Secretary, but with a succession of different titles, from 1956 to 1977. Until 1965, when Blackwell Scientific Publications Ltd. were appointed as publishers of the Society's journals he was responsible, under the direction of the Secretaries, for the Society's publications. This meant that, in addition to overseeing the Society's activities in general, he subedited all of the papers and corresponded with authors, referees and printers. When the growth of the journals led to the appointment of Blackwell Scientific Publications as publishers, he continued to handle all of the correspondence with authors, referees and editors until S.G. Sykes was appointed as an Editorial Assistant in 1973. During his time with the Society Rubidge saw a great increase in its membership and in the scope of the Society's scientific activities and he also handled the upheaval brought about by the conversion of the Society's old Meeting Room in Burlington House into offices and a Fellows' Room. The Society was an essential part of his life and he worked very long hours and the officers found it very difficult to persuade him to take holidays. His departure was noted by a presentation to him at the 1977 December Ordinary

Meeting and in 1978 a tribute to him appeared in *Quarterly Journal*. The presentation was not from the Society as such but from those Fellows who had served on Council, Council's committees or Editorial Boards during his time as Executive Secretary.

Although it was going to be very difficult to follow Edward Rubidge, Alan Hume did so admirably. His background was very different from that of any previous Executive Secretary. He was a career serviceman who had just taken early retirement from the Royal Air Force as a Group Captain at the age of 48 in order to make a second career. He had considerable administrative experience, both in command of operational units and in staff posts. Although he had previously been responsible for very large numbers of subordinate staff, he seemed equally at home in the more intimate surroundings of the Society. While carrying on the work of Edward Rubidge, he rapidly made his own mark on the running of the Society's office. To quote three examples: he made the arrangement for variable covenants with the Inland Revenue which has already been mentioned; he also persuaded officers and Council that the redecoration and maintenance of the Society's premises should be planned on a ten-year cycle and that a reserve fund should be set up to which contributions would be made each year (the intention was that very large and expensive jobs would not have to be deferred because expenditure in any one year would produce a deficit in the Society's accounts); he also followed up earlier work by Treasurers of the Society with the aim of ensuring that the Society made an economic charge to other users of the Society's premises including the British Astronomical Association and the London Mathematical Society. Alan Hume obviously wished to work for the Society until a normal retirement age and it was therefore a great shock to officers and Council and to the many friends that he had already made in the Society when he died after a sudden and totally unexpected heart attack in 1979 December, just short of two years in office.

There followed a period without an Executive Secretary until 1980 May, during which time Mrs Enid Lake, the Librarian, doubled as Acting General Secretary. Then during the final eight months of the decade the Executive Secretary was Mr J.A. (Tony) Steff-Langston. He was also a recently retired Group Captain who had, in fact, been a colleague and friend of Alan Hume. The time to describe his period as Executive Secretary will be in a future volume of the *History* but here it should be said that at the end of 1980 the running of the Society appeared to be in very firm and able hands.

In the first half of the decade there was concern that the Society did not have a really adequate pension scheme for its staff. At the same time there was a succession of changes, actual or projected, in the state pensions scheme. Because of the uncertainty in the state scheme and, in particular, in the probable provisions for contracting out of the scheme, there was delay in the introduction of a better Society scheme. This was introduced on a provisional basis in 1975 January but the rules were not finally confirmed until 1977 May when the rules of the new state scheme, to be introduced in 1978 April, were known. As a result the scheme was fully operational when Edward Rubidge retired and when Alan Hume died.

The Society's Premises

In the late 1960s there had been a major transformation in the Society's premises with the conversion of the Meeting Room into the Fellows' Room and offices and with the redecoration of the Council Room. In addition, when the Royal Society moved from Burlington House to its new accommodation in Carlton House Terrace, the consequent changes in Burlington House led to the Society being allocated some additional rooms on the other side of the courtyard. At the start of the decade the decorative state of the remainder of the Society's premises was poor, there was an urgent need for rewiring and the solid fuel central heating boiler was close to the end of its useful life. At the same time the state of the Society's finances was poor. As a result there were several years in which the Honorary Auditors in their report to the Annual General Meeting pointed out serious deficiencies in the premises and in particular they were concerned with the decorative condition of the caretakers' flat and of the general office.

As the decade advanced the Society did manage to carry out all of the necessary work including the renewal of the wiring, the installation of a new heating system, the decoration of the hall, staircase and landing, of the general office and the Assistant Secretary's office and of the Grove-Hills and Spencer Rooms. When there was a change of caretakers in 1977, an opportunity was taken to completely modernize their accommodation. In 1979 it was discovered that both the inner entrance doors and the library doors needed to be replaced at considerable expense. Council decided that it would be appropriate to use the recent bequest from A.D. Thackeray for the library doors and subsequently there was a formal opening when members of his family were present.

Much of the work in the premises, particularly in the second half of the decade, was inspired by Dr Valerie Myerscough who became Convener of the House Committee in 1974 and who was also a Secretary from 1979 until her early death in 1980. The major work of the decade with which she was particularly concerned was the restoration of a room on the first floor. This room had for a long time been divided into two, part being a ladies' toilet and the other part being a general storage or junk room. Because of internal partitions it was not clear that it was potentially a handsome room comparable with the Council Room on the floor below. A new ladies' toilet was installed at the end of a corridor and the whole room was opened out and restored. It was named the Herschel Room and it was formally opened in 1979 by Caroline Herschel a Fellow of the Society and a direct descendant of Sir William and Sir John Herschel. Plate 13 is a photograph taken on that occasion. The room was decorated with the Society's portraits of William, Caroline (William's sister) and John Herschel. The Society thus ended the decade with the whole premises in good decorative order and with a rolling programme aimed at doing necessary work in rotation.

The Library

The decade proved to be a very important one for the library, although there were occasions when financial pressure on the Society caused both the Library Committee and Council to have very serious worries about its future role. It had become clear that the Society could no longer hope to have a comprehensive collection of journals and text-books on astronomy, let alone build up a similar collection in geophysics. In addition university libraries were, in general, better equipped than they had been in earlier decades and a much smaller proportion of Fellows made regular use of the library. There was therefore some temptation to suggest that it should cease to be a contemporary working library and that it should become a historical library. This idea in turn raised two problems.

The first was that it was asked why Fellows should be expected to support the library from their Annual Contributions if only historians would make much use of it. Indeed at one Annual General Meeting a Fellow seriously suggested that the library was a millstone around the neck of the Society and that the Society should offer the library to some other body that would be better placed to preserve it, and at the same time this would make possible a reduction in the Annual Contribution of Fellows. This did not receive sufficient support to be seriously considered, although officers did learn of one other learned society which had handed its library over to the British Library in return for a preferential access for the Society's Fellows.

The second problem raised by the idea of having a historical library was that it would only fulfil that function at the end of the 21st century, say, if it contained the best books published in the 20th century. As history does not stop at some prescribed date, a continuing process of accession is needed. Although it might be possible to fill some gaps once the importance of a book was recognized, past experience suggested that an attempt should be made to decide which books would

Plate 13. Opening of the Herschel Room. 1979 February 8. Left to right N. Le Grand (Royal Society). M.J. Seaton (President). Miss Caroline Herschel. Dr J.R. Shakeshaft. Dr V.P. Myerscough. Mr D.G. Savage (Editorial Assistant). A.J. Hume (Executive Secretary). (© M.J. Smyth.)

be influential and to buy them as they were published. It was eventually decided that the Society would try to acquire the most important mainstream astronomical and geophysical journals and that it would be very selective in the acquisition of books and particularly of conference proceedings. It was also decided to stop the exchange of the Society's publications with most institutions which did not provide useful material for the library. This was of substantial advantage to the Society and little cost to the library. Council agreed that special consideration should continue to be given to institutions with serious foreign exchange problems. As in the past, Fellows would be encouraged to present copies of their own books to the Society. It was also suggested that the library might benefit if members bequeathed books to the Society but here it was recognized that there would be a danger that the Society might acquire too many duplicate copies.

The Herschel archive

Probably the most important development of the decade was the cataloguing of the Herschel archive. For many years a box of Herschel papers had been lying uncatalogued at the Society's premises but in 1974 Council was persuaded that this very important source of astronomical history must be rescued and catalogued, and that this was so important that a temporary appointment of an archival assistant should be made for two years to see the work through. This was a brave decision soon after the Annual Contribution had been doubled. After advertisement the post was offered to Dr J.A. Bennett, who took up his appointment in 1974 October. The appointment was a great success and Bennett made many important discoveries. As has already been mentioned, the final issue of *Memoirs* was a catalogue of the Herschel archive and microfilm copies of the archive were sold to many institutions around the world.

The Historical Collection

The other main event of the decade was the progress made with the rebinding of the Society's pre-1800 books and with their housing. A grant had previously been obtained from the Pilgrim Trust to help with the rebinding of the Society's oldest books, many of which were in a precarious state. In the second half of the decade additional grants were obtained from the Radcliffe Trustees, the Pilgrim Trust and the British Library, with the expectation that all of the pre-1800 books would be rebound by the early 1980s. The British Library also made a grant for the publication of the catalogue of the Herschel archive. The existence of these grants towards the up-keep of the Society's historical collection made it much easier for Council to justify the general expenditure on the library. In 1975 July an important decision was taken to glaze the library shelves in the Grove-Hills and Spencer Rooms which contain the Society's oldest and most valuable books. The main purpose of the glazing was security as theft of important books would be made more difficult. It also provided some additional protection for the books. A year later Council agreed that it must in future have a realistic insurance cover for the whole contents of the library. Although many of the old books would be irreplaceable in the event of fire or theft, it was clear the library was the Society's largest asset and that this asset must be properly safeguarded.

The Librarians

The Librarian at the start of the decade was R.E.W. Maddison, who worked about three days a week and who was helped by a Library Assistant or Assistant Librarian. He was not a professional librarian but he was a scientist and scholar who had a great fund of knowledge about books and journals. He served the Society very well. He had been appointed as Librarian after he had retired from other employment and his appointment was neither for a fixed term nor was a retirement age attached to it. In the middle of the decade he had difficulty carrying out his duties on a regular basis because of illness of his wife and in 1975 he agreed to retire and was succeeded by the Assistant Librarian Mrs E. Lake. Officers hoped that Maddison would agree to continue as a consultant and to do some of the work for which his skills were particularly valuable but he decided to retire completely and he was given an ex gratia pension. He immediately took the opportunity to become a Fellow of the Society, which he could not be while he was an employee, and he continued to be a fairly regular visitor to the Society's premises. Mrs Lake was a qualified librarian and the advantage of having a Librarian who was on duty full-time was soon recognized. She was a very successful Librarian and, as has already been mentioned, she was a very efficient acting General Secretary following the untimely death of Alan Hume. The Library Committee and particularly its Chairman was very valuable both to the Librarian and to Council and the Society was well-served by two very able committee Chairmen, G.J. Whitrow and D.W. Dewhirst.

The Society's Instrument Collection

Several times during the decade an attempt was made to sort out the Society's instrument collection. This had been started in 1828 to provide serviceable instruments for loan to Fellows. Over the years the important historical items in the collection had been placed on permanent loan to observatories and national museums and some more immediately useful items had been given to the British Astronomical Association. In 1973 April it was agreed that the most valuable instrument, the Harrison Clock, should be transferred from the Royal Greenwich Observatory to the National Maritime Museum and the following year it was agreed that the Museum would insure it on the basis of regular valuations by Sothebys. In 1977 permission was given for the Museum to rehouse the clock at its own expense. Plate 14 is a recent photograph of the clock at Greenwich. In 1975 October an *ad hoc* Instruments Committee was set up to make recommendations relating to the disposal of instruments and equipment on the Society's premises. It was intended that some further long-term loans might be made to museums and that, in addition, all existing loan agreements should be updated. The Council received powers at the General meeting of 1979 October to dispose of the remaining items of historical interest either by gift or sale or by placing them on permanent loan with appropriate institutions but the final resolution was left to the 1980s.

Education in Astronomy

One body associated with the Society which had very mixed fortunes during the decade was the Education Committee. The committee was set up on an informal basis in 1973 and its structure was formalized in 1974. Because of the great public interest in astronomy, it was felt that the Society should take an interest in astronomical education in the widest possible sense. Although it was a committee of Council, it was not concerned with education in geophysics.

One activity which was not started by the Education Committee but with which it became associated was the Young People's Lecture. The idea for such a lecture was suggested by McNally in 1971 May. He hoped that the Society could sponsor a lecture that would be given to a live audience and would be simultaneously televised. In 1972 February he was able to interest the Open University in the lecture and the first lecture was given by Dr D.W. Sciama in 1973 November. The lecture on black holes was given at St Andrews and shown on BBC television. It was hoped that such lectures would be given biennially and in 1974 December Professor A. Hewish was selected as the second lecturer. Early hopes were not realized, mainly because of financial problems. The Society could only make a limited contribution from its trust funds and the Open University was not in such a good financial position as when the idea had first been proposed. The lecture by Hewish, on pulsars, was finally given in 1978 February. On this occasion there was no live audience and it was a studio production. Following that lecture,

Professor McCrea approached IBM UK Ltd for support for future lectures and later in 1978 a grant was obtained which enabled a third lecture by Dr G.E. Hunt on exploration of the planets to be given in 1980 February. It must be emphasized that even on this occasion the major fraction of the cost was borne by the Open University/BBC.

The booklet on postgraduate opportunities in astronomy has already been mentioned. This was proposed by Dr S. Mitton and the Education Committee was only indirectly concerned with it. A venture of the committee itself was the production of an information booklet on astronomy for the general public. It was intended that this could be sold by organizations such as the Royal Observatories and Jodrell Bank, which had exhibitions, and that university departments could send them out in response to routine enquiries about astronomy. The booklet contained a general description of the Universe with illustrations, a reading list and hints about obtaining further information about astronomy. The first edition was ready for sale in early 1978.

The Education Committee's largest venture was also ultimately its least successful. At the time the committee was being set up it was suggested that the Society should involve itself with the teaching of astronomy in schools which it was felt might increase if the curriculum was broadened. The suggestion was that the Society should concern itself with the training of teachers in mathematics and physics who might find themselves able to introduce some astronomy into their teaching. The committee decided that it would like to produce course units for use in colleges of education and the Society succeeded in obtaining a grant from the Nuffield Foundation to finance the preparatory work. A considerable amount of work was done by members of the Education Committee and its consultants but unfortunately the original plan for the units became impracticable for two main reasons. The first was that changes in teacher training targets and the closure of colleges of education resulting from the reactions of government to the falling birth rate led to some reduction of interest in the course units in colleges. In addition inflation made a very large increase in the estimated cost of producing the units. The Education Committee was unable to find a publisher who would be willing to take over production of the units including the financial risk. At the same time it was clear that the Society could not itself undertake such a risky venture. Ultimately it was decided that the project must be abandoned.

Early in 1980 the Education Committee and the Society became involved in another publishing venture which involved no financial risk to the Society. Mitchell Beazley Publications were producing an *Atlas of the Solar System*, which was in fact a series of books about objects in the Solar System. They asked the Society to be associated with the publication in the sense that it would provide an expert to vet the texts

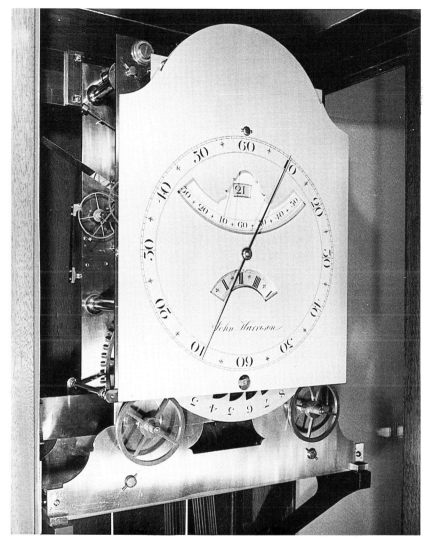

Plate 14. The Harrison Clock at the National Maritime Museum, Greenwich. (© National Maritime Museum.)

of the books before publication and would allow the series to be called a Royal Astronomical Society series. In return the Society received royalties and a fee for the expert who checked the manuscripts. As a significant advance royalty was payable, this was a useful input to the Society's funds once it was satisfied that the series would be respectable.

The Royal Astronomical Society Club

Although the RAS Club is completely distinct from the Society, its members must all be members of the Society and the relations between the Club and the Society are close. This became particularly true during the 1970s when an informal arrangement was started whereby the Club would invite distinguished guests of the Society to dine at the Club in return for the Society paying for the guest's dinner. As a result many medallists and George Darwin or Harold Jeffreys Lecturers were able to enjoy a relaxing evening with astronomical and geophysical colleagues without special arrangements having to be made. The Club dinners which were held at out-of-town meetings in Manchester, York, Durham and Cardiff provided an ideal opportunity for the Society as well as the Club to say thank you to the hosts at the meetings. Current members of the Society's Council are always entitled to dine at meetings of the Club, other than the Annual Parish Dinner in January, and such ex-officio dining has often been a prelude to election to the Club.

The Observatory

It seems appropriate to end this chapter with reference to another body whose close connection with the Society still continues. The reports of the Society's Ordinary Meetings in *The Observatory* constitute the *Hansard* of astronomy, even if the cut and thrust of question and answer is sometimes polished between the meeting and publication. Only *The Observatory* provides any permanent record of what actually happened at the meetings. Readers to that journal are, however, asked to think twice before believing the account of the meeting which was reported in its 1000th issue at which the President was D. Inkwell and the Secretaries R.J. Tinker and J.R. Crankshaft and where the Bronze Medal was due to be presented to Dr R. Gryphon. The Society owes a debt to successive editors of *The Observatory* and especially to Dr R.F. Griffin, who was an Editor throughout the decade, for their faithful record of the Society's activities. Answers to Council's questionnaire to members in 1975 indicated that most members, including many geophysicists, were very pleased to receive *The Observatory* on a regular basis.

The Future

When early in its history the Society adopted the motto 'Quicquid nitet notandum', astronomy consisted of visual observations with the aid of the telescope and the motto announced that members of the Society were ready to study anything which emitted light. During the first century of the Society's history these visual observations were supplemented by photography and spectroscopy but there remained only the slightest indication that useful information about the Universe would be carried in other parts of the electromagnetic spectrum. The period covered by this volume has seen astronomers studying objects that shine in all parts of the electromagnetic spectrum from long radio waves to gamma rays and also the start of the search for gravitational radiation. Astronomers are also concerned with particle radiation, starting with cosmic rays and continuing with solar neutrinos, and with the possibility of detecting other elementary particles which may contribute much of the mass density in the Universe. The improvement in detectors has been so great that optical astronomers are now looking to even larger telescopes as the only means of penetrating deeper into space. Not long before the start of the new astronomies the view was expressed in a popular astronomy text that no new types of object remained to be discovered. Despite the advances that have been made, few astronomers would express the same view today. As we enter the closing decades of the Society's second century I think that we can be confident that the motto 'Quicquid nitet notandum' is as appropriate as ever and that those Fellows who celebrate the second centenary will have many new discoveries to applaud.

APPENDIX 1
OFFICERS OF THE SOCIETY
1920–80

Presidents

Titles and reference to published version of Presidential Addresses are given: if the address was on the award of a Gold Medal, the name of the medallist is given in brackets.

1919–21   A. Fowler
     1920   The early history of the Society (*Mon. Not. R. astr. Soc.*, **80**, 440)
     1921   Contributions to the study of stellar evolution (H.N. Russell) (*Mon. Not. R. astr. Soc.*, **81**, 334)
1921–23   A.S. Eddington
     1922   Theories of cosmogony (J.H. Jeans) (*Mon. Not. R. astr. Soc.*, **82**, 279)
     1923   Applications of the interferometer to astronomical measurements (A.A. Michelson) (*Mon. Not. R. astr. Soc.*, **83**, 309)
1923–25   J.L.E. Dreyer
     1924   On the desirability of publishing a new edition of Isaac Newton's collected works (*Mon. Not. R. astr. Soc.*, **84**, 298)
     1925   No formal address
1925–27   J.H. Jeans
     1926   Relativity and the theory of gravitation (A. Einstein) (*Mon. Not. R. astr. Soc.*, **86**, 262)
     1927   Stellar Parallax and astronomical photography (F. Schlesinger) (*Mon. Not. R. astr. Soc.*, **87**, 340)
1927–29   T.E.R. Phillips
     1928   Theory of the four great satellites of Jupiter (R.A. Sampson) (*Mon. Not. R. astr. Soc.*, **88**, 334)
     1929   Determination of the distance of the lesser Magellanic cloud (E. Hertzsprung) (*Mon. Not. R. astr. Soc.*, **89**, 404)
1929–31   A.C.D. Crommelin
     1930   Observations of stellar radial velocities and important conclusions derived from them (J.S. Plaskett) (*Mon. Not. R. astr. Soc.*, **90**, 466)
     1931   Theoretical investigations on the orbits of the satellites of Jupiter *and* the theory of relativity (W. de Sitter) (*Mon. Not. R. astr. Soc.*, **91**, 422)
1931–33   H. Knox-Shaw
     1932   Double stars (R.G. Aitken) (*Mon. Not. R. astr. Soc.*, **92**, 354)
     1933   The distances and motions of the extra-galactic nebulae (*Mon. Not. R. astr. Soc.*, **93**, 304)
1933–35   F.J.M. Stratton
     1934   International co-operation in astronomy: a chapter of astronomical history (*Mon. Not. R. astr. Soc.*, **94**, 361)
     1935   Radiative equilibrium and theory of stellar atmospheres (E.A. Milne) (*Mon. Not. R. astr. Soc.*, **95**, 421)

1935–37 J.H. Reynolds
  1936 The galactic nebulae (*Mon. Not. R. astr. Soc*, **96**, 393)
  1937 Physics of the Earth and other planets and the study of the origin and age of the solar system (H. Jeffreys) (*Mon. Not. R. astr. Soc.*, **97**, 375)
1937–39 H. Spencer Jones
  1938 Spectra of gaseous nebulae and of novae *and* the photography of planets in light of different colours (W.H. Wright) (*Mon. Not. R. astr. Soc.*, **98**, 358)
  1939 Some problems of meridian astronomy (*Mon. Not. R. astr. Soc.*, **99**, 424)
1939–41 H.C. Plummer
  1940 Distances, velocities, distribution and nature of the extragalactic nebulae (E.P. Hubble) (*Mon. Not. R. astr. Soc.*, **100**, 342)
  1941 The development of the vertical telescope (*Mon. Not. R. astr. Soc.*, **101**, 165)
1941–43 S. Chapman
  1942 The source of the Sun's energy (*Mon. Not. R. astr. Soc.*, **102**, 110)
  1943 Magnetism in the Earth's atmosphere (*Mon. Not. R. astr. Soc.*, **103**, 117)
1943–45 E.A. Milne
  1944 On the nature of universal gravitation (*Mon. Not. R. astr. Soc.*, **104**, 120)
  1945 The natural philosophy of stellar structure (*Mon. Not. R. astr. Soc.*, **105**, 146)
1945–47 H.H. Plaskett
  1946 Astronomical telescopes (*Mon. Not. R. astr. Soc.*, **106**, 80)
  1947 Astronomical spectroscopy (*Mon. Not. R. astr. Soc.*, **107**, 117)
1947–49 W.M.H. Greaves
  1948 The photometry of the continuous spectrum (*Mon. Not. R. astr. Soc.*, **108**, 131)
  1949 Geophysics and solar physics, particularly geomagnetic phenomena (S. Chapman) (*Mon. Not. R. astr. Soc.*, **109**, 258)
1949–51 W.M. Smart
  1950 Physical methods in astronomical spectroscopy (J Stebbins) (*Mon. Not. R. astr. Soc.*, **110**, 179)
  1951 The first half of the twentieth century: a partial review (*Mon. Not. R. astr. Soc.*, **111**, 247)
1951–53 H. Dingle
  1952 Stellar parallaxes and the problems of star positions and stellar motions (J. Jackson) (*Mon. Not. R. astr. Soc.*, **112**, 345)
  1953 On science and modern cosmology (*Mon. Not. R. astr. Soc.*, **113**, 393)
1953–55 J. Jackson
  1954 Observations of galactic and extragalactic objects (W. Baade) (*Mon. Not. R. astr. Soc.*, **114**, 370)
  1955 On the need for observations in the older branches of astronomy, especially double stars (*Mon. Not. R. astr. Soc.*, **115**, 204)
1955–57 Sir Harold Jeffreys
  1956 The Earth's thermal history (*Mon. Not. R. astr. Soc.*, **116**, 231)
  1957 Probability theory in astronomy (*Mon. Not. R. astr. Soc.*, **117**, 347)

1957–59 W.H. Steavenson
1958 Astronomical photometry, fundamental astronomy and the design of astronomical instruments (A. Danjon) (*Mon. Not. R. astr. Soc.*, **118**, 401)
1959 The instruments of Sir William Herschel (*Mon. Not. R. astr. Soc.*, **119**, 449)
1959–61 R.O. Redman
1960 The work of the Cambridge observatories (*Q. Jl R. astr. Soc.*, **1**, 10)
1961 Photometry in astronomy (*Q. Jl R. astr. Soc.*, **2**, 96)
1961–63 W.H. McCrea
1962 Evidence of evolution in astronomy (*Q. Jl R. astr. Soc.*, **3**, 63)
1963 Cosmology, a brief review (*Q. Jl R. astr. Soc.*, **4**, 185)
1963–65 Sir Richard Woolley
1964 Observations in the southern hemisphere (*Q. Jl R. astr. Soc.* **5**, 110)
1965 The R R Lyrae variable stars (*Q. Jl R. astr. Soc.*, **6**, 125)
1965–67 T.G. Cowling
1966 The development of the theory of stellar structure (*Q Jl R. astr. Soc.*, **7**, 121)
1967 Interstellar and interplanetary plasmas (*Q. Jl R. astr. Soc.*, **8**, 130)
1967–69 D.H. Sadler
1968 Astronomical measures of time (*Q. Jl R. astr. Soc.*, **9**, 281)
1969 Astronomy and navigation (*Q. Jl R. astr. Soc.*, **10**, 97)
1969–71 Sir Bernard Lovell
1970 The prospects for British astronomy (*Q. Jl R. astr. Soc.*, **11**, 71)
1971 Sporadic outbursts of red dwarf stars (*Q. Jl R. astr. Soc.*, **12**, 98)
1971–73 Sir Fred Hoyle
1972 The history of the Earth (*Q. Jl R. astr. Soc.*, **13**, 328)
1973 The origin of the Universe (*Q. Jl R. astr. Soc.*, **14**, 278)
1973–75 D.E. Blackwell
1974 Stellar atmospheres and laboratory astrophysics (*Q. Jl R. astr. Soc.*, **15**, 224)
1975 Uncertainty in astronomy (*Q. Jl R. astr. Soc.*, **16**, 361)
1975–77 F. Graham Smith
1976 How do pulsars pulse (*Q. Jl R. astr. Soc.*, **17**, 383)
1977 Telescopes and instrumentation for the northern hemisphere observatory (not published)
1977–79 A.H. Cook
1978 Physics of celestial masers (*Q. Jl R. astr. Soc.*, **19**, 255)
1979 Geophysics and the human condition (*Q. Jl R. astr. Soc.*, **20**, 229)
1979– M.J. Seaton
1980 Spectra of gaseous nebulae (*Q. Jl R. astr. Soc.*, **21**, 229)

Other addresses on awards of medals delivered by Presidents. The medal, medallists and reference are given.

J.L.E. Dreyer   Jackson Gwilt. A.S. Williams and W.S. Franks (*Mon. Not. R. astr. Soc.*, **83**, 428)
                   Gold. A.S. Eddington (*Mon. Not. R. astr. Soc.*, **84**, 548)

J.H. Jeans   Gold. F.W. Dyson (*Mon. Not. R. astr. Soc.*, **85**, 672)

T.E.R.Phillips	Jackson Gwilt. W.H. Steavenson and W. Reid (*Mon. Not. R. astr. Soc.*, **88**, 344)
A.C.D. Crommelin	Jackson Gwilt. C.W. Tombaugh (*Mon. Not. R. astr. Soc.*, **91**, 434)
H. Knox-Shaw	Gold. V.M. Slipher (*Mon. Not. R. astr. Soc.*, **93**, 476)
F.J.M. Stratton	Gold. H. Shapley (*Mon. Not. R. astr. Soc.*, **94**, 592) Jackson Gwilt. W.F. Gale (*Mon. Not. R. astr. Soc.*, **95**, 429)
J.H. Reynolds	Gold. H. Kimura (*Mon. Not. R. astr. Soc.*, **96**, 390)
H. Spencer Jones	Jackson Gwilt. F.J. Hargreaves and P.M. Ryves (*Mon. Not. R. astr. Soc.*, **98**, 375)
H.C. Plummer	Gold. B. Lyot (*Mon. Not. R. astr. Soc.*, **99**, 538)
S. Chapman	Jackson Gwilt. R.L. Waterfield (*Mon. Not. R. astr. Soc.*, **102**, 41) Gold. H. Spencer Jones (*Mon. Not. R. astr. Soc.*, **103**, 116)
E.A. Milne	Gold. O. Struve (*Mon. Not. R. astr. Soc*, **104**, 112) Gold. B Edlén (*Mon. Not. R. astr. Soc.*, **105**, 138)
H.H. Plaskett	Jackson Gwilt. H.W. Newton (*Mon. Not. R. astr. Soc.*, **106**, 2) Gold. J.H. Oort (*Mon. Not. R. astr. Soc.*, **106**, 242)
W.M.H. Greaves	Gold. M.G.J. Minnaert (*Mon. Not. R. astr. Soc.*, **107**, 243) Gold. B. Lindblad (*Mon. Not. R. astr. Soc.*, **108**, 212) Jackson Gwilt. A.M. Newbigin (*Mon. Not. R. astr. Soc.*, **109**, 118)
W.M. Smart	Gold. A. Pannekoek (*Mon. Not. R. astr. Soc.*, **111**, 245)
H. Dingle	Gold. S. Chandrasekhar (*Mon. Not. R. astr. Soc.*, **113**, 270) Eddington. G. Lemaître (*Mon. Not. R. astr. Soc.*, **113**, 271) Jackson Gwilt. J.P.M. Prentice (*Mon. Not. R. astr. Soc.*, **113**, 272)
J. Jackson	Gold. D. Brouwer (*Mon. Not. R. astr. Soc.*, **115**, 199) Eddington. H.C. van de Hulst (*Mon. Not. R. astr. Soc.*, **115**, 202)
Sir Harold Jeffreys	Gold. T.G. Cowling (*Mon. Not. R. astr. Soc.*, **116**, 229) Gold. A Unsöld (*Mon. Not. R. astr. Soc.*, **117**, 344) Jackson Gwilt. R.P. de Kock (*Mon. Not. R. astr. Soc.*, **117**, 346)
W.H. Steavenson	Eddington. H.W. Babcock (*Mon. Not. R. astr. Soc.*, **118**, 404) Gold. R.A. Lyttleton (*Mon. Not. R. astr. Soc.*, **119**, 445) Eddington. J.S. Hey (*Mon. Not. R. astr. Soc.*, **119**, 447)

R.O. Redman	Gold. V.A. Ambartsumian (*Q. Jl R. astr. Soc.*, **1**, 23) Eddington. R. d'E. Atkinson (*Q. Jl R. astr. Soc.*, **1**, 26) Jackson Gwilt. F.M. Bateson and A.F.A.L. Jones (*Q. Jl R. astr. Soc.*, **1**, 27) Eddington. H.A. Bethe (*Q. Jl R. astr. Soc.*, **2**, 107) Gold. H. Zanstra (*Q. Jl R. astr. Soc.*, **2**, 109)
W.H. McCrea	Gold. B Strömgren (*Q. Jl R. astr. Soc.*, **3**, 81) Eddington. A. Lallemand (*Q. Jl R. astr. Soc.*, **3**, 84) Gold. H.H. Plaskett (*Q. Jl R. astr. Soc.*, **4**, 176) Eddington. A.R. Sandage and M. Schwarzschild (*Q. Jl R. astr. Soc.*, **4**, 180) Jackson Gwilt. G.E.D. Alcock (*Q. Jl R. astr. Soc.*, **4**, 183)
Sir Richard Woolley	Gold. M. Ryle (*Q. Jl R. astr. Soc.*, **5**, 103) Gold. M. Ewing (*Q. Jl R. astr. Soc.*, **5**, 106) Eddington. H. Friedman and R. Tousey (*Q. Jl R. astr. Soc.*, **5**, 108) Gold. Sir Edward Bullard (*Q. Jl R. astr. Soc.*, **6**, 118) Gold. G.M. Clemence (*Q. Jl R. astr. Soc.*, **6**, 120) Eddington. R.V. Pound and G.A. Rebka (*Q. Jl R. astr. Soc.*, **6**, 123)
T.G. Cowling	Gold. I.S. Bowen (*Q. Jl R. astr. Soc.*, **7**, 114) Gold. H.C. Urey (*Q. Jl R. astr. Soc.*, **7**, 117) Eddington. R. Wildt (*Q. Jl R. astr. Soc.*, **7**, 120) Gold. H. Alfvén (*Q. Jl R. astr. Soc.*, **8**, 123) Gold. A.R. Sandage (*Q. Jl R. astr. Soc.*, **8**, 126) Eddington. R.F. Christy (*Q. Jl R. astr. Soc.*, **8**, 129)
D.H. Sadler	Gold. F. Hoyle (*Q. Jl R. astr. Soc.*, **9**, 271) Gold. W.H. Munk (*Q. Jl R. astr. Soc.*, **9**, 274) Eddington. R. Hanbury Brown and R.Q. Twiss (*Q. Jl R. astr. Soc.*, **9**, 277) Jackson Gwilt. J.G. Porter (*Q. Jl R. astr. Soc.*, **9**, 279) Gold. A.T. Price (*Q. Jl R. astr. Soc.*, **10**, 89) Gold. M. Schwarzschild (*Q. Jl R. astr. Soc.*, **10**, 91) Eddington. A. Hewish (*Q. Jl R. astr. Soc.*, **10**, 95)
Sir Bernard Lovell	Gold. H.W. Babcock (*Q. Jl R. astr. Soc.*, **11**, 85) Eddington. C. Hayashi (*Q. Jl R. astr. Soc.*, **11**, 88) Gold. F. Press (*Q. Jl R. astr. Soc.*, **12**, 133) Gold. Sir Richard Woolley (*Q. Jl R. astr. Soc.*, **12**, 135) Eddington. D.G. King-Hele (*Q. Jl R. astr. Soc.*, **12**, 138) Jackson Gwilt. A.W.J. Cousins (*Q. Jl R. astr. Soc.*, **12**, 139)
Sir Fred Hoyle	Gold. F. Zwicky (*Q. Jl R. astr. Soc.*, **13**, 483) Gold. H.I.S Thirlaway (*Q. Jl R. astr. Soc.*, **13**, 484) Eddington. P. Ledoux (*Q. Jl R. astr. Soc.*, **13**, 485) Gold. F. Birch (*Q. Jl R. astr. Soc.*, **14**, 274) Gold. E.E. Salpeter (*Q. Jl R. astr. Soc.*, **14**, 275) Chapman. D.H. Matthews and F.J. Vine (*Q. Jl R. astr. Soc.*, **14**, 276)

D.E. Blackwell   Gold. L. Biermann (*Q. Jl R. astr. Soc.*, **15**, 219)
                       Gold. K. Bullen (*Q. Jl R. astr. Soc.*, **15**, 220)
                       Herschel. P. Wild (*Q. Jl R. astr. Soc.*, **15**, 221)
                       Jackson Gwilt. G.E. Perry (*Q. Jl R. astr. Soc.*, **15**, 222)
                       Gold. J.L. Greenstein (*Q. Jl R. astr. Soc.*, **16**, 356)
                       Gold. E.J. Öpik (*Q. Jl R. astr. Soc.*, **16**, 358)
                       Eddington. S.W. Hawking and R. Penrose (*Q. Jl R. astr. Soc.*, **16**, 359)

F. Graham Smith   Gold. J.A. Ratcliffe (*Q. Jl R. astr. Soc.*, **17**, 214)
                       Gold. W.H. McCrea (*Q. Jl R. astr. Soc.*, **17**, 215)

A.H. Cook   Gold. D.R. Bates (*Q. Jl R. astr. Soc.*, **19**, 153) (delivered by F. Graham Smith)
                       Gold. J.G. Bolton (*Q. Jl R. astr. Soc.*, **19**, 153)
                       Herschel. A.A. Penzias and R.W. Wilson (*Q. Jl R. astr. Soc.*, **19**, 156)
                       Gold. L. Spitzer (*Q. Jl R. astr. Soc.*, **20**, 1)
                       Eddington. W.A. Fowler (*Q. Jl R. astr. Soc.*, **20**, 3)

M.J. Seaton   Chapman. E. Parker (*Q. Jl R. astr. Soc.*, **21**, 73)
                       Gold. C.L. Pekeris (*Q. Jl R. astr. Soc.*, **21**, 245)
                       Gold. L. Knopoff (*Q. Jl R. astr. Soc.*, **22**, 1)

Obituary notices of former Presidents who died in the period 1920–80. The Presidents are listed in the order of their election. The names of the authors of obituaries are given in brackets. Where initials are shown first, it is these that accompany the notice.

J.W.L. Glaisher (1848–1928) *Mon. Not. R. astr. Soc.*, **89**, 300 (HHT, H.H. Turner)
Sir William H.M. Christie (1845–1922) *Mon. Not. R. astr. Soc.*, **83**, 233 (HPH, H.P. Hollis)
E.B. Knobel (1841–1930) *Mon. Not. R. astr. Soc.*, **91**, 318 (FWD, F.W. Dyson)
Sir W. de W. Abney (1843–1920) *Mon. Not. R. astr. Soc.*, **81**, 250 (HPH. H.P. Hollis)
H.H. Turner (1861–1930) *Mon. Not. R. astr. Soc.*, **91**, 321 (HCP, H.C. Plummer)
W.H. Maw (1838–1924) *Mon. Not. R. astr. Soc.*, **85**, 311 (EBK, E.B. Knobel)
H.F. Newall (1857–1944) *Mon. Not. R. astr. Soc.*, **105**, 95 (F.J.M. Stratton)
Sir Frank Dyson (1868–1939) *Mon. Not. R. astr. Soc.*, **100**, 238 (J. Jackson)
E.H. Grove-Hills (1864–1922) *Mon. Not. R. astr. Soc.*, **83**, 241 (anon)
R.A. Sampson (1866–1939) *Mon. Not. R. astr. Soc.*, **100**, 258 (W.H.M. Greaves)
P.A. MacMahon (1854–1929) *Mon. Not. R. astr. Soc.*, **90**, 373 (HHT, H.H. Turner)
A. Fowler (1868–1940) *Mon. Not. R. astr. Soc.*, **101**, 132 (H.C. Plummer)
Sir Arthur Eddington (1882–1944) *Mon. Not. R. astr. Soc.*, **105**, 68 (H. Spencer Jones, E.T. Whittaker)
J.L.E. Dreyer (1852–1926) *Mon. Not. R. astr. Soc.*, **87**, 251 (EBK, E.B. Knobel)
Sir James Jeans (1877–1946) *Mon. Not. R. astr. Soc.*, **107**, 46 (E.A. Milne)
T.E.R. Phillips (1868–1942) *Mon. Not. R. astr. Soc.*, **103**, 70 (W.H. Steavenson)
A.C.D. Crommelin (1869–1939) *Mon. Not. R. astr. Soc.*, **100**, 234 (C.R. Davidson)

H. Knox-Shaw (1885–1970) *Q. Jl R. astr. Soc.*, **12**, 197 (A.D. Thackeray)
F.J.M. Stratton (1881–1960) *Q. Jl R. astr. Soc.*, **2**, 44 (R.v.d. R. Woolley)
J.H. Reynolds (1874–1949) *Mon. Not. R. astr. Soc.*, **110**, 131 (M. Johnson)
Sir Harold Spencer Jones (1890–1960) *Q. Jl R. astr. Soc.*, **4**, 113 (D.H. Sadler)
H.C. Plummer (1875–1946) *Mon. Not. R. astr. Soc.*, **107**, 56 (W.M. Smart)
S.Chapman (1888–1970) *Q. Jl R. astr. Soc.*, **13**, 464 (T.G. Cowling, V.C.A. Ferraro)
E.A. Milne (1896–1950) *Mon. Not. R. astr. Soc.*, **111**, 160 (W.H. McCrea, H.H. Plaskett)
H.H. Plaskett (1893–1980) *Q. Jl R. astr. Soc.*, **21**, 486 (M.G. Adam)
W.M.H. Greaves (1897–1955) *Mon. Not. R. astr. Soc.*, **116**, 145 (J. Jackson)
W.M. Smart (1889–1975) *Q. Jl R. astr. Soc.*, **18**, 140 (M.W. Ovenden)
H. Dingle (1890–1978) *Q. Jl R. astr. Soc.*, **21**, 333 (G.J. Whitrow)
J. Jackson (1887–1958) *Mon. Not. R. astr. Soc.*, **119**, 345 (H. Spencer Jones)
W.H. Steavenson (1894–1975) *Q. Jl R. astr. Soc.*, **18**, 147 (D.W. Dewhirst)
R.O. Redman (1905–1975) *Q. Jl R. astr. Soc.*, **17**, 80 (D.E. Blackwell, D.W. Dewhirst)

Treasurers

1913–22	E.B. Knobel	1937–46	J.H. Reynolds
1922	E.H. Grove–Hills	1946–52	Sir H. Spencer Jones
1923–27	F.J.M. Stratton	1952–56	L.M. Milne–Thomson
1927–28	P.H. Hepburn	1956–67	G. Merton
1928–29	E.B. Knobel	1967–76	A. Hunter
1929–35	J.H. Reynolds	1976–79	W.H. McCrea
1935–37	Sir F.W. Dyson	1979–	R.J. Tayler

Secretaries

1917–23	A.C.D. Crommelin	1949–54	F.M. McBain
1919–26	T.E.R. Phillips	1949–56	A. Hunter
1923–24	H. Spencer Jones	1954–57	C.W. Allen
1924–29	J. Jackson	1956–64	H. Bondi
1926–30	H. Knox-Shaw	1957–66	M.W. Ovenden
1929–33	H. Dingle	1964–71	F. Graham Smith
1930–37	W.M. Smart	1964–73	T.F. Gaskell
1933–39	W.M.H. Greaves	1966–72	D. McNally
1937–40	H.H. Plaskett	1971–79	R.J. Tayler
1939–47	D.H. Sadler	1972–78	J.R. Shakeshaft
1940–41	R.d'E. Atkinson	1973–78	J.A. Hudson
1941–43	A.D. Thackeray	1978–	R.D. Davies
1943–46	H.R. Hulme	1978–	J.A. Jacobs
1946–49	W.H. McCrea	1979–80	V.P. Myerscough
1947–48	A.D. Thackeray	1980–	C. Jordan
1948–49	R.O. Redman		

Foreign Secretaries

1919–30	H.H. Turner	1936–44	Sir Arthur Eddington
1931–36	A. Fowler	1945–55	F.J.M. Stratton

1955—60 Sir Harold Spencer Jones   1962—64   C.W. Allen
1961—62   R.O. Redman

The office of Foreign Secretary ceased to exist in 1964 from which date the third Secretary of the Society was a geophysicist. From 1964 to 1978 there was a Foreign Correspondent, who was not an officer.

Foreign Correspondents

1964—68	C.W. Allen		1973—75	B.E.J. Pagel
1968—71	W.H. McCrea		1975—77	C.A. Murray
1971—73	R.H. Stoy		1977—78	D. Walsh

APPENDIX 2

SENIOR MEMBERS

OF THE SOCIETY'S STAFF

Assistant Secretaries

1875–1922	W.H. Wesley
1923–43	Miss K. Williams
1943–47	Miss E.M. Wadsworth
1947–56	Miss M.F.M. Garratt
1956–77	E.C. Rubidge (title changed to General Secretary in 1963 and to Executive Secretary in 1977)

Executive Secretaries

1978–79	A.J. Hume
1980–	J.A. Steff-Langston

Mr Rubidge was Editorial Superintendent in 1956 January to March. Mrs E. Lake was Acting General Secretary from 1979 December to 1980 May.

Librarians

1948–55	Miss E.M. Wadsworth
1955–57	Mrs M.H. Markiewicz
1957–60	Mrs J.E. Phillips
1960–65	Mrs B. Harrington
1965–75	Dr R.E.W. Maddison
1975–	Mrs E. Lake

Before 1948 there was no formal post of Librarian, the Assistant Secretary having direct responsibility for the library.

APPENDIX 3

THE SOCIETY'S MEDALLISTS

1920–80

Gold Medal

1921	H.N. Russell	1960	V.A. Ambartsumian
1922	J.H. Jeans	1961	H. Zanstra
1923	A.A. Michelson	1962	B. Strömgren
1924	A.S. Eddington	1963	H.H. Plaskett
1925	Sir F.W. Dyson	1964	M. Ryle
1926	A. Einstein		M. Ewing
1927	F. Schlesinger	1965	E. Bullard
1928	R.A. Sampson		G.M. Clemence
1929	E. Hertzsprung	1966	I.S. Bowen
1930	J.S. Plaskett		H.C. Urey
1931	W. de Sitter	1967	H. Alfvén
1932	R.G. Aitken		A.R. Sandage
1933	V.M. Slipher	1968	F. Hoyle
1934	H. Shapley		W. Munk
1935	E.A. Milne	1969	A.T. Price
1936	H. Kimura		M. Schwarzschild
1937	H. Jeffreys	1970	H.W. Babcock
1938	W.H. Wright	1971	F. Press
1939	B. Lyot		R.v.d. R. Woolley
1940	E.P. Hubble	1972	H.I.S. Thirlaway
1943	H. Spencer Jones		F. Zwicky
1944	O. Struve	1973	F. Birch
1945	B. Edlén		E.E. Salpeter
1946	J. H. Oort	1974	L. Biermann
1947	M.G.J. Minnaert		K.E. Bullen
1948	B. Lindblad	1975	J.L. Greenstein
1949	S. Chapman		E.J. Öpik
1950	J. Stebbins	1976	W.H. McCrea
1951	A. Pannekoek		J.A. Ratcliffe
1952	J. Jackson	1977	D.R. Bates
1953	S. Chandrasekhar		J.G. Bolton
1954	W. Baade	1978	L. Spitzer
1955	D. Brouwer		J.A. van Allen
1956	T.G. Cowling	1979	L. Knopoff
1957	A. Unsöld		C.G. Wynne
1958	A. Danjon	1980	C.L. Pekeris
1959	R.A. Lyttleton		M. Schmidt

Eddington Medal

1953	G. Lemaître	1966	R. Wildt
1955	H.C. van de Hulst	1967	R.F. Christy
1958	H.W. Babcock	1968	R. Hanbury Brown
1959	J.S. Hey		R.Q. Twiss
1960	R.d'E. Atkinson	1969	A Hewish
1961	H.A. Bethe	1970	C. Hayashi
1962	A. Lallemand	1971	D.G. King-Hele
1963	A.R. Sandage	1972	P. Ledoux
	M. Schwarzschild	1975	S. Hawking
1964	H. Friedman		R. Penrose
	R. Tousey	1978	W.A. Fowler
1965	R.V. Pound		
	G.A. Rebka		

Chapman Medal

1973	D.H. Matthews	1976	S−I. Akasofu
	F.J. Vine	1979	E.N. Parker

Herschel Medal

1974	J.P. Wild	1980	G. de Vaucouleurs
1977	A.A. Penzias		
	R.W. Wilson		

The Hannah Jackson (née Gwilt) Gift and Medal

1923	A.S. Williams	1953	J.P.M. Prentice
	W.S. Franks	1956	R.P. de Kock
1928	W. Reid	1960	F.M. Bateson
	W.H. Steavenson		A.F.A.L. Jones
1931	C.W. Tombaugh	1963	G.E.D. Alcock
1935	W.F. Gale	1968	J.G. Porter
1938	F.J. Hargreaves	1971	A.W.J. Cousins
	P.M. Ryves	1974	G.E. Perry
1942	R.L. Waterfield	1977	P.A. Moore
1946	H.W. Newton	1980	R.F. Griffin
1949	A.M. Newbigin		

APPENDIX 4

THE SOCIETY'S NAMED LECTURES

George Darwin Lecturers 1927–80

The name of the lecturer, the title and date of the lecture, the venue if outside London and a reference to the published version are given.

1927   F. Schlesinger. Some aspects of astronomical photometry of precision. 1927 March 11 (*Mon. Not. R. astr. Soc.*, **87**, 506)

1928   W.H. Wright. On photographs of the brighter planets by light of different colours. 1928 June 8 (*Mon. Not. R. astr. Soc.*, **88**, 709)

1929   E. Hertzsprung. The Pleiades. 1929 May 10 (*Mon. Not. R. astr. Soc.*, **89**, 660)

1930   J.S. Plaskett. The high-temperature stars. 1930 May 9 (*Mon. Not. R. astr. Soc.*, **90**, 616)

1931   W. de Sitter. Jupiter's Galilean satellites. 1931 May 8 (*Mon. Not. R. astr. Soc.*, **91**, 706)

1932   R.G. Aitken. What we know about double stars. 1932 May 13 (*Mon. Not. R. astr. Soc.*, **92**, 596)

1933   V.M. Slipher. Spectrographic studies of the planets. 1933 May 12 (*Mon. Not. R. astr. Soc.*, **93**, 657)

1934   H. Shapley. On some structural features of the metagalaxy. 1934 May 11 (*Mon. Not. R. astr. Soc.*, **94**, 791)

1935   H.N. Russell. The analysis of spectra and its applications in astronomy. 1935 June 14 (*Mon. Not. R. astr. Soc.*, **95**, 610)

1936   A. Kopff. Star catalogues, especially those of fundamental character. 1936 June 10 (*Mon. Not. R. astr. Soc.*, **96**, 714)

1937   N.E. Nörlund. Astronomical longitude and azimuth determination. 1937 May 14 (*Mon. Not. R. astr. Soc.*, **97**, 489)

1938   C. Fabry. Interstellar space. 1938 May 13 (*Mon. Not. R. astr. Soc.*, **98**, 681)

1939   B. Lyot. A study of the solar corona and prominences without eclipses. 1939 May 12 (*Mon. Not. R. astr. Soc.*, **99**, 580)

1940–42   No lecture

1943   S. Rosseland. The pulsation theory of Cepheid variables. 1943 September 10 (*Mon. Not. R. astr. Soc.*, **103**, 233)

1944   J. Proudman. The tides of the Atlantic ocean. 1944 October 13, (*Mon. Not. R. astr. Soc.*, **104**, 244)

1945   B. Edlén. Identification of coronal lines. 1945 October 12 (*Mon. Not. R. astr. Soc.*, **105**, 323)

1946   J.H. Oort. Some phenomena connected with interstellar matter. 1946 May 10 (*Mon. Not. R. astr. Soc.*, **106**, 159)

1947   M.G.J. Minnaert. The Fraunhofer lines of the solar spectrum. 1947 May 9 (*Mon. Not. R. astr. Soc.*, **107**, 274)

1948 B. Lindblad. On the dynamics of stellar systems. 1948 April 9 (*Mon. Not. R. astr. Soc.*, **108**, 214)

1949 O. Struve. Spectroscopic binaries. 1949 October 14 (*Mon. Not. R. astr. Soc.*, **109**, 487)

1950 J. Stebbins. The electrical photometry of stars and nebulae. 1950 October 13 (*Mon. Not. R. astr. Soc.*, **110**, 416)

1951 A. Pannekoek. The origin of astronomy. 1951 April 13 (*Mon. Not. R. astr. Soc.*, **111**, 347)

1952 E.P. Hubble. The law of red-shifts. 1953 May 8 (*Mon. Not. R. astr. Soc.*, **113**, 658)

1953 S. Chandrasekhar. Problems of stability in hydrodynamics and hydro-magnetics. 1953 November 13 (*Mon. Not. R. astr. Soc.*, **113**, 667)

1954 No lecture

1955 D. Brouwer. The motions of the outer planets. 1955 April 6 (*Mon. Not. R. astr. Soc.*, **115**, 221)

1956 W. Baade. Stellar populations. 1957 May 10 (no full version published, see *The Observatory* **77**, 117)

1957 A. Unsöld. On the quantitative analysis of stellar spectra. 1957 November 8 (*Mon. Not. R. astr. Soc.*, **118**, 3)

1958 A. Danjon. The contribution of the impersonal astrolabe to fundamental astronomy. 1958 May 9 (*Mon. Not. R. astr. Soc.*, **118**, 411)

1959 A.A. Mikhailov. The deflection of light by the gravitational field of the Sun. 1959 April 10 (*Mon. Not. R. astr. Soc.*, **119**, 593)

1960 V.A. Ambartsumian. On the evolution of stellar systems. 1960 May 13 (*Q. Jl R. astr. Soc.*, **1**, 152)

1961 H. Zanstra. The gaseous nebula as a quantum counter. 1961 October 13 (*Q. Jl R. astr. Soc.*, **2**, 137)

1962 B. Strömgren. Problems of internal constitution and kinematics of main sequence stars. 1962 October 12 (*Q. Jl R. astr. Soc.*, **4**, 8)

1963 R. Tousey. The spectrum of the Sun in the extreme ultra-violet. 1963 October 11 (*Q. Jl R. astr. Soc.*, **5**, 123)

1964 P. Swings. Cometary spectra. 1964 October 9 (*Q. Jl R. astr. Soc.*, **6**, 28)

1965 G.M. Clemence. Inertial frames of reference. 1965 October 8 (*Q. Jl R. astr. Soc.*, **7**, 10)

1966 I.S. Bowen. Future tools of the astronomer. 1966 October 14 (*Q. Jl R. astr. Soc.*, **8**, 9)

1967 R.F. Christy. The theory of Cepheid variability. 1967 September 8 (*Q. Jl R. astr. Soc.*, **9**, 13)

1968 F. Hoyle. Highly condensed objects. 1968 October 11 (*Q. Jl R. astr. Soc.*, **10**, 10)

1969 M. Schwarzschild. Stellar evolution in globular clusters. 1969 October 10 (*Q. Jl R. astr. Soc.*, **11**, 12)

1970 H.C. van de Hulst. Cosmic ray electrons. 1970 November 13 (*Q. Jl R. astr. Soc.*, **13**, 10)

1971 O. Gingerich. Johannes Kepler and the new astronomy. 1971 December 10 (*Q. Jl R. astr. Soc.*, **13**, 346)

1972 P. Connes. Astronomical Fourier spectroscopy. 1972 October 13 (*Q. Jl R. astr. Soc.*, **14**, 288)

1973 W.A. Fowler. High temperature nuclear astrophysics. 1973 December 14 (*Q. Jl R. astr. Soc.*, **15**, 82)

1974   V.L. Ginzburg. Does astronomy need new physics? 1975 April 11 (delivered by M.S. Longair) (*Q. Jl R. astr. Soc.*, **16**, 265)

1975   L. Spitzer. Hydrogen molecules in interstellar space. 1975 December 12 (*Q. Jl R. astr. Soc.*, **17**, 97)

1976   M.J. Rees. Quasars and young galaxies. 1976 April 8 (Manchester) (*Q. Jl R. astr. Soc.*, **18**, 429)

1977   H. van der Laan. Radio galaxies: their structure and violent nuclei. 1977 April 5 (Edinburgh) (no published version)

1978   L. Goldberg. Some problems connected with mass loss in red giant stars. 1979 May 11 (*Q. Jl R. astr. Soc.*, **20**, 361)

1979   I.R. King. The dynamics of globular clusters. 1979 October 12 (*Q. Jl R. astr. Soc.*, **22**, 227)

1980   J.H. Taylor. Gravitational waves and the binary pulsar. 1980 May 9 (no full version published, see *The Observatory*, **100**, 143)

Harold Jeffreys Lecturers 1963–80

1963   Sir Harold Jeffreys. How soft is the Earth? 1963 October 25 (*Q. Jl R. astr. Soc.*, **5**, 10)

1964   M. Ewing. The sediments of the Argentine basin. 1964 October 23 (*Q. Jl R. astr. Soc.*, **6**, 10)

1965   Sir Edward Bullard. Electromagnetic induction in the Earth. 1965 December 3 (*Q. Jl R. astr. Soc.*, **8**, 143)

1966   H.C. Urey. The abundance of the elements with special reference to the problem of the iron abundance. 1966 September 23 (*Q. Jl R. astr. Soc.*, **8**, 23)

1967   H. Alfvén. On the origin of the solar system. 1967 May 12 (*Q. Jl R. astr. Soc.*, **8**, 215)

1968   W. Munk. Once again — tidal friction. 1968 March 29 (*Q. Jl R. astr. Soc.*, **9**, 352)

1969   A.T. Price. The electrical conductivity of the Earth. 1969 October 24 (*Q. Jl R. astr. Soc.*, **11**, 23)

1970   F. Press. The Earth and the Moon. 1971 March 12 (*Q. Jl R. astr. Soc.*, **12**, 232)

1971   D.G. King-Hele. Heavenly harmony and earthly harmonics. 1971 December 10 (*Q. Jl R. astr. Soc.*, **13**, 374)

1972   H.I.S. Thirlaway. Forensic seismology. 1972 December 8 (*Q. Jl R. astr. Soc.*, **14**, 297)

1973   A. Dollfus. Remote study of the solar system bodies by optical polarimetry. 1973 October 12 (no published version)

1974   J.W. Dungey. Some remaining mysteries in the aurora. 1974 November 8 (*Q. Jl R. astr. Soc*, **16**, 117)

1975   D.P. McKenzie. Plate tectonics and its driving mechanism. 1976 May 14 (no published version)

1976   L. Knopoff. Lateral inhomogeneities in the Earth's mantle. 1976 November 12 (no published version)

1977   J.W. King. The influence of solar phenomena and weather and climate. 1977 November 11 (no full version published, see *The Observatory*, **98**, 81)

1978   M.M. Woolfson. Cosmogony today. 1978 October 13 (*Q. Jl R. astr. Soc.*, **20**, 97)

1979   C. Sagan. The exploration of the outer solar system. 1979 April 11 (Durham) (no published version)

1980   G.J. Wasserburg. Galactic nucleosynthesis and the early history of the solar system. 1981 March 13 (no published version)

APPENDIX 5

STATISTICAL INFORMATION

1920−80

This appendix contains six figures giving information about the membership, publications and finances of the Society for the period 1920−80. Notes on each figure follow in place of figure captions.

Figure 1: Membership

It can be seen from this figure that, following an increase immediately after the First World War, there was very little change in membership between 1920 and 1950. The drop in membership during the Second World War was minimized because arrangements were made to keep on the Society's books members who were temporarily unable to pay their dues. The membership increased steadily from 1950 until the early 1970s with a particularly rapid increase starting in 1959. The membership was steady again in the 1970s with temporary falls in membership being related to large increases in Annual Contribution. The membership at the end of the period was probably effectively the largest ever because after 1977 only those members who were up-to-date with their dues were included, whereas earlier it was possible to remain on the list until two years' contributions were owed.

The Annual Contribution remained at its original (1820) figure until 1947. Only in the 1970s was there any very substantial increase.

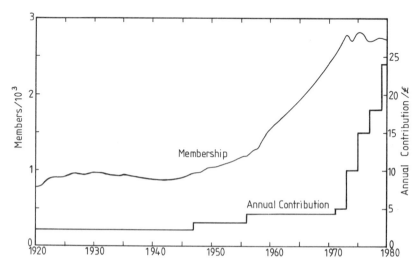

Figure 2: Publications

In this figure is shown the total number of pages published by the Society in its journals: *Monthly Notices; Memoirs; Geophysical Supplements; Occasional Notes; Geophysical Journal*, and *Quarterly Journal*. The figures for different years are not precisely comparable because there were some changes in page size and in the size and density of type. However, the general character is clear. There was little change in the number of pages between 1920 and 1940. There was a substantial reduction in and shortly after the Second World War due both to absence of astronomers on war service and to restrictions on the use of paper. After 1950 the increase in pages published paralleled the increase in number of members.

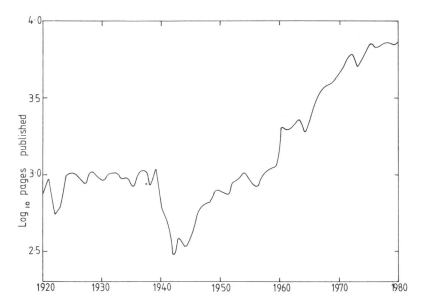

Figures 3–6: Finances

Figure 3

The incomes from members and from investments are shown for periods for
which it is easy to extract these figures from the published accounts. The
extremely rapid increase in investment income towards the end of the period is
apparent. The drop in income from members during the Second World War
indicates what fraction of the membership was unable to pay dues.

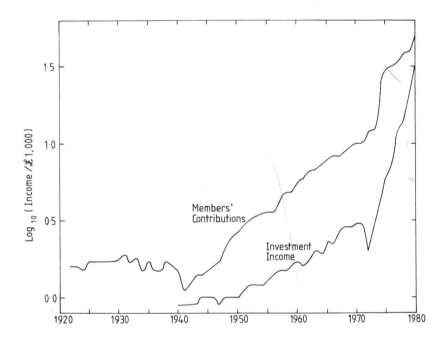

Figure 4

The book value of the Society's investments (reserves) and the annual turnover of the Society's accounts are shown. Although fluctuations in market value are much greater than those in book value, as has already been shown in Table 2, over a long period the book value gives a fair indication of the trend of the reserves. It is notable that until the late 1950s the reserves were several times greater than the turnover but that by 1970 the opposite was true. This goes some way to explain why the financial crisis of the 1920s could be faced without an increase in Annual Contribution, whereas the Annual Contribution increased by a factor of more than five between 1970 and 1980.

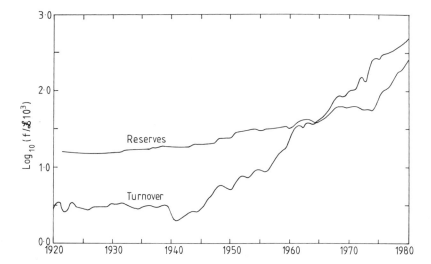

Figure 5

The surplus or deficit on the annual accounts is shown as a percentage of turnover. It can be seen that there have been very rapid fluctuations, some of which can be accounted for by special non-recurrent items of income or expenditure.

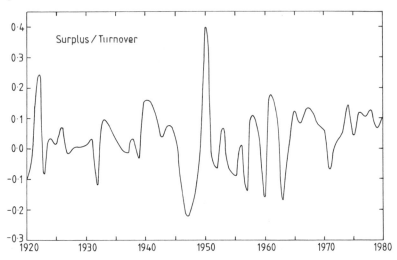

Figure 6

The fluctuations in Figure 5 have been smoothed by taking five year averages; each point represents an average over the year in question and the two on either side. This average curve shows that the Society's accounts were close to balance from 1920 to 1940. A surplus built up during the Second World War when expenditure was restricted but this was followed by a deficit when full activities were resumed. There was then a period of fluctuations in which there were problems in adjusting to a rapid increase in membership and publications. The surplus rose following the appointment of Blackwell Scientific Publications Ltd as the Society's publishers. The surpluses of the 1970s were largely necessary to maintain the value of the Society's investments in real terms during a period of rapid inflation.

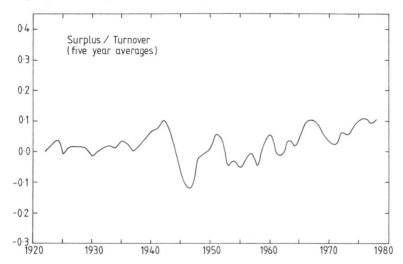

Surplus / Turnover (five year averages)

INDEX

As explained in the Introduction, names are not entered in the index unless they are associated with some particular event which is described. In particular the election and retirement of officers, awards of medals and delivery of named lectures are not normally indexed because there are complete lists in the Appendices. Numbers in bold face printed at the end of an entry refer to plates.